Chemical Engineering Primer with Computer Applications

T0143610

Chemical Engineering Primer with Computer Applications

Hussein K. Abdel-Aal

CRC Press is an imprint of the
Taylor & Francis Group, an **informa** business

MATLAB® is a trademark of The MathWorks, Inc. and is used with permission. The MathWorks does not warrant the accuracy of the text or exercises in this book. This book's use or discussion of MATLAB® software or related products does not constitute endorsement or sponsorship by The MathWorks of a particular pedagogical approach or particular use of the MATLAB® software.

CRC Press
Taylor & Francis Group
6000 Broken Sound Parkway NW, Suite 300
Boca Raton, FL 33487-2742

© 2017 by Taylor & Francis Group, LLC
CRC Press is an imprint of Taylor & Francis Group, an Informa business

No claim to original U.S. Government works

Printed on acid-free paper
Version Date: 20160822

International Standard Book Number-13: 978-1-4987-3057-0 (Paperback)

This book contains information obtained from authentic and highly regarded sources. Reasonable efforts have been made to publish reliable data and information, but the author and publisher cannot assume responsibility for the validity of all materials or the consequences of their use. The authors and publishers have attempted to trace the copyright holders of all material reproduced in this publication and apologize to copyright holders if permission to publish in this form has not been obtained. If any copyright material has not been acknowledged please write and let us know so we may rectify in any future reprint.

Except as permitted under U.S. Copyright Law, no part of this book may be reprinted, reproduced, transmitted, or utilized in any form by any electronic, mechanical, or other means, now known or hereafter invented, including photocopying, microfilming, and recording, or in any information storage or retrieval system, without written permission from the publishers.

For permission to photocopy or use material electronically from this work, please access www.copyright.com (http://www.copyright.com/) or contact the Copyright Clearance Center, Inc. (CCC), 222 Rosewood Drive, Danvers, MA 01923, 978-750-8400. CCC is a not-for-profit organization that provides licenses and registration for a variety of users. For organizations that have been granted a photocopy license by the CCC, a separate system of payment has been arranged.

Trademark Notice: Product or corporate names may be trademarks or registered trademarks, and are used only for identification and explanation without intent to infringe.

Library of Congress Cataloging-in-Publication Data

Names: Abdel-Aal, Hussein K., author.
Title: Chemical engineering primer with computer applications / Hussein K. Abdel-Aal.
Description: Boca Raton : CRC Press, 2017. | Includes bibliographical references.
Identifiers: LCCN 2016034556 | ISBN 9781498730570 (paperback : acid-free paper)
Subjects: LCSH: Chemical engineering--Data processing.
Classification: LCC TP184 .A24 2017 | DDC 660--dc23
LC record available at https://lccn.loc.gov/2016034556

Visit the Taylor & Francis Web site at
http://www.taylorandfrancis.com

and the CRC Press Web site at
http://www.crcpress.com

Printed and bound in the United States of America by Publishers Graphics, LLC on sustainably sourced paper.

Contents

Preface ... xiii
Acknowledgments ... xvii
Author ... xix
Introduction: Initial Thoughts ... xxi

SECTION I An Insight into Chemical Engineering

Chapter 1 Introductory Concepts ... 3

1.1 System of Units ... 3
1.2 Process Variables ... 6
1.2.1 Temperature Scales ... 6
1.2.2 Pressure Scale ... 7
1.3 Gas Laws ... 9
1.3.1 Boyle's Law ... 9
1.3.2 Charles' Law ... 9
1.3.3 Dalton's Law ... 9
1.3.4 Ideal Gas Law ... 10
1.3.5 Raoult's Law ... 12
1.3.6 Amagat's Law ... 13
1.3.7 Avogadro's Hypothesis ... 14
End-of-Chapter Solved Examples ... 14
References ... 18

Chapter 2 Basic Principles and Introduction to Calculations ... 19

2.1 The Two Building Blocks of Chemical Engineering: Unit Operation and Unit Processes ... 19
2.2 Chemical Process Industry ... 20
2.3 Chemical Plant ... 21
2.4 Process Design ... 22
2.5 Role of Material and Energy Balances, Thermodynamics, and Kinetics ... 22
2.6 Design Variables for Sizing Piping and Process Equipment ... 23
2.7 Basic Definitions ... 24
2.8 Conversion of Physical Events and Principles to Mathematical Formula ... 24
2.9 Basic Laws ... 25
2.9.1 Mass and Energy Conservation: Material Balance and Heat Balance ... 25

2.9.2 Thermodynamics......26
2.9.3 Some General Tips......26
2.9.3.1 *Parts-Per* Notation......26
2.9.3.2 Frequently Asked Questions......27
End-of-Chapter Solved Examples......27
Reference......32

SECTION II Fundamentals and Problem-Solving Profile

Chapter 3 Numerical Methods and Chemical Engineering Computations......35
3.1 Introduction......35
3.2 Basic Definitions and Introductory Remarks......36
3.2.1 Algorithms......36
3.2.2 Programming Language......36
3.3 Areas and Domain of Numerical Methods......36
3.4 Applications Using Elementary Numerical Methods......37
End-of-Chapter Solved Examples......38
References......43

Chapter 4 The Approach to Solve Problems by Computers......45
4.1 Introduction......45
4.2 Methodology......46
4.3 Model Development and Mathematical Formulation......47
4.4 Applications......50
4.4.1 Problem Statement......50
End-of-Chapter Solved Examples......52
References......65

SECTION III Backbone Materials

Chapter 5 Fluid Flow and Transport of Fluids......69
5.1 Introduction......69
5.2 Definitions and Terminology in Fluid Mechanics......70
5.2.1 Compressible and Noncompressible Fluids......70
5.2.2 Pressure Concept......70
5.2.3 Forces Acting on a Fluid......70
5.2.4 Nature of Fluids......71

5.3 Classification of Fluid Flow ... 71
5.3.1 Ideal Fluid Flow ... 72
5.3.2 Real Fluid Flow ... 72
5.3.3 Steady and Unsteady State Flow ... 72
5.3.4 Uniform and Nonuniform Flow ... 72
5.3.5 Laminar Flow ... 72
5.3.6 Turbulent Flow ... 72
5.4 Parameters in Laminar Flow ... 73
5.4.1 Velocity Gradient or Rate of Shear Stress (du/dy) ... 73
5.4.2 Shear Stress (τ) ... 74
5.5 Fluid Statics ... 74
5.5.1 Gauge Pressure and Absolute Pressure ... 74
5.6 Overall Energy Balance Equation ... 75
5.6.1 Introduction ... 75
5.6.2 General Energy Balance Equation ... 75
5.6.3 Special Cases ... 76
5.6.4 Mechanical Energy Balance ... 76
5.6.5 Bernoulli's Equation ... 77
5.6.6 Head Form of Bernoulli's Equation ... 78
5.6.7 Pressure Drop and Friction Losses ... 78
5.7 Piping and Pumps ... 80
5.7.1 Introduction ... 80
5.7.2 Piping ... 80
5.7.2.1 Sizing of Pipelines ... 80
5.7.2.2 Economic Balance in Piping and Optimum Pipe Diameter ... 80
5.7.2.3 Stepwise Procedure to Calculate the Wall Thickness ... 81
5.7.2.4 Relationship between the Pipe Diameter and the Pressure Drops ... 82
5.7.3 Pumps ... 83
5.7.3.1 Introduction ... 83
5.7.3.2 Classification and Types of Pumps ... 84
5.7.3.3 How to Select a Pump? ... 85
5.7.3.4 Calculation of the Horsepower for a Pump ... 86
End-of-Chapter Solved Examples ... 86
References ... 94

Chapter 6 Heat Transmission ... 95

6.1 Introduction ... 95
6.2 Modes of Heat Transfer ... 96
6.3 Heat Transfer by Conduction ... 97

6.3.1 Fourier's Law 97
6.3.2 Thermal Conductivity 98
6.3.3 Conduction through a Hollow Cylinder 99
6.3.4 Conduction through Solids in Series 99
6.3.5 Solved Examples 99
6.4 Heat Transfer by Convection 101
6.4.1 Introduction 101
6.4.2 Methodology of Calculation 102
6.4.3 Convective Heat Transfer Coefficient 102
6.5 Overall Heat Transfer Coefficient 103
6.5.1 Background 103
6.5.2 Discussion 103
6.6 Heat Transfer by Radiation 106
6.6.1 Introduction 106
6.6.2 Blackbody Radiation 107
6.7 Heat Exchangers 107
6.7.1 Types of Heat Exchangers 107
6.7.2 Log Mean Temperature Difference 108
6.7.3 Design of Heat Exchangers 109
End-of-Chapter Solved Examples 110
6A Appendix: Internal-Flow Convection Correlations 116
Reference 117

Chapter 7 Two-Phase Gas–Liquid Diffusional Operations: Distillation and Absorption 119

7.1 Part I: Distillation 120
7.1.1 Introduction 120
7.1.1.1 Basic Concepts and Principles 120
7.1.2 Three Pillars for Distillation 123
7.1.3 Features of Distillation Units 124
7.1.3.1 Types of Distillation Columns 125
7.1.3.2 Overhead Condensers 125
7.1.3.3 Reflux and Reflux Ratio: Top of Column 126
7.1.3.4 Reboilers: Bottom of Column 127
7.1.3.5 Rectification and Stripping Sections in Distillation Columns 127
7.1.3.6 Effect of Operating Pressure 127
7.1.4 Design of Distillation Columns 129
7.1.4.1 Introduction 129
7.1.4.2 Approach to Solve Distillation Problems 130
7.1.4.3 Physical Models for Distillation 131
7.1.4.4 Calculation of Number of Trays 131

7.1.5 McCabe–Thiele Method 132
7.1.5.1 Introduction 132
7.1.5.2 Assumptions 132
7.1.5.3 Procedure 132
7.1.5.4 Illustration of How to Use Excel to Solve a Problem 134
7.1.5.5 Total and Minimum Reflux: Limiting Cases 138
7.1.5.6 Tray Efficiency 138
7.1.5.7 Stage Design and Efficiency 140
7.1.6 Shortcut Methods: Fenske–Underwood–Gilliland 142
7.1.6.1 Introduction 142
7.1.6.2 Fenske Equation 142
7.1.6.3 Underwood Equation 142
7.1.6.4 Gilliland's Method 143
7.1.7 Column Diameter 144
7.1.8 Flash Distillation 144
7.1.8.1 Flash Equation 146
7.1.9 Reactive Distillation 147
7.2 Part II: Absorption and Stripping 147
7.2.1 Basic Concepts 148
7.2.2 Outline of Design Approach 149
7.2.3 Graphical Methods: Solution for Simple Countercurrent Absorption Operations 150
7.2.4 Material Balance Calculations 151
7.2.5 Analytical Methods: Equations to Calculate the Number of Stages for Absorbers/Strippers 154
7.2.6 Transfer Unit Concept for Absorption in Packed Columns 156
7.2.7 Role of Absorption in Natural Gas Field Processing 156
End-of-Chapter Solved Examples 157
Reference 164

Chapter 8 Reaction Kinetics, Chemical Reactors, and Thermodynamics 165

8.1 Part I: Fundamentals of Reaction Kinetics 166
8.1.1 Introduction 166
8.1.2 Six Categories of Chemical Reactions 167
8.1.3 Reaction Rates 168
8.1.3.1 Forms of Rate Law 168
8.2 Part II: Types and Design of Chemical Reactors 172
8.2.1 Introduction 172
8.2.2 Types of Laboratory and Industrial Catalytic Reactors 173

8.2.3 Catalytic Reactors 174
8.2.4 Design of Chemical Reactors 175
8.2.4.1 Basic Design Equations 176
8.3 Part III: Introduction to Thermodynamics 178
8.3.1 Basic Definitions and Concepts 178
8.3.1.1 The Concept of Equilibrium 178
8.3.1.2 The Concept of a Process 178
8.3.1.3 Equations of State 178
8.3.2 Thermodynamic Laws and Entropy 179
8.3.2.1 Entropy 179
8.3.3 Role of ΔG in Chemical Reactions 180
End-of-Chapter Solved Examples 181
Part (a): Manual Calculations 181
Part (b): Using MATLAB® 184
References 186

Chapter 9 Chemical Plant Design and Process Economics 187

9.1 Introduction 188
9.2 Plant Design is the Heart of Chemical Engineering 188
9.3 How to Handle a Design Project? 189
9.3.1 Proposed Procedure 189
9.3.2 A–Z Chart for Plant Design 190
9.3.3 Flow Sheeting and Types of Flow Diagrams 191
9.3.4 Scale-Up Practice and Safety Factors 191
9.4 Other Aspects in Plant Design 191
9.5 The Role of Economics in Plant Design 193
9.5.1 Introduction 193
9.5.2 Estimation of the Fixed Capital Investment and the Total Capital Investment 193
9.5.3 Decision Making Use of Annuity 194
9.5.4 Profitability Analysis Using Economic Indicators 195
9.5.4.1 The Annual Rate of Return 195
9.5.4.2 Payout Period, Payback Time, or Cash Recovery Period 196
9.5.4.3 Discounted Cash Flow Rate of Return and Present Value Index 196
9.5.4.4 Net Present Value 196
End-of-Chapter Solved Examples 196

Chapter 10 Case Studies 201

10.1 Introduction 201
10.2 Case Study 1: Alternatives to Storage of Ammonia 202
10.2.1 Introduction 202
10.2.2 The Problem 202

10.2.3 Approach ... 202
10.2.4 Conclusions ... 203
10.2.5 Discussions ... 204
10.3 Case Study 2: Flash Dewatering of Raw Sewage Effluents ... 204
10.3.1 Introduction ... 204
10.3.2 The Problem ... 204
10.3.3 Proposed Scheme ... 204
10.3.4 Process Description ... 206
10.3.5 Conclusions ... 207
10.3.6 Discussions ... 208
10.4 Case Study 3: Magnesium Extraction from Seawater ... 208
10.4.1 Introduction ... 208
10.4.2 Proposed Method ... 209
10.4.3 Production of Magnesium Metal from Magnesium Chloride ... 210
10.4.4 Discussions ... 210
10.5 Case Study 4: Chemical Desalting of Brines ... 210
10.5.1 Introduction ... 210
10.5.2 Methodology ... 210
10.5.3 Main Reactions ... 211
10.5.3.1 Process Synthesis with Modifications ... 212
10.5.3.2 Consumption–Production Analysis ... 212
10.5.3.3 Comparison between the Solvay Process and the Proposed Process ... 213
10.5.4 Conclusion ... 215
10.5.5 Discussion and Problem Formulation ... 215
10.6 Case Study 5: Applications of Differential Equations ... 216
10.6.1 Introduction ... 216
10.6.2 Problem Statement ... 216
10.6.3 Proposed Approach ... 216
10.6.4 Discussions ... 218
10.7 Case Study 6: Reactive Distillation ... 218
10.7.1 Introduction ... 218
10.7.2 Objective ... 218
10.7.3 Discussion ... 219
References ... 219

Appendix A: Application Problems ... 221

Appendix B: Mathematical Methods and Review Notes ... 235

Appendix C: Conversion Tables ... 243

Bibliography ... 247

Index ... 251

Preface

It has been observed that most of the textbooks written for introductory courses in chemical engineering follow a formal classical method. This applies, in particular, to texts for sophomore courses. Authors normally focus on writing many chapters on material balance and energy balance followed by those on allied topics. What we have used here is a different approach to offer an introductory course in chemical engineering, with emphasis on handling computer solutions for chemical engineering problems. Our aim is to introduce the reader to chemical engineering fundamentals without the distractions, which is the case with the material found in many textbooks.

This book has been written as a *beginning* chemical engineering text with emphasis on *computer applications.* The word *beginning* is the underlying paradigm of writing this book. It is the act or *process of bringing* this knowledge to young students majoring in chemical engineering at different levels. The meaning may also infer the *beginning to walk into the real world* right after finishing the bachelor's degree. In this way, this book should be of use to both chemical engineering students as a point of departure for their entry into industry and the practicing engineer for their continuing education program.

For a practicing chemical engineer, whether in industry or teaching, this book would be a good companion that adequately treats chemical engineering as a composite and integrated field.

This book covers the core concepts of chemical engineering, ranging from conservation laws all the way up to chemical kinetics. For nonengineers, who are expected to work with chemical engineers on projects, scale-ups, and process evaluations, this *primer* would be an excellent source for a solid understanding of basic concepts of chemical engineering analysis, design, and calculations.

The text is divided into three sections.

SECTION I: INSIGHT INTO CHEMICAL ENGINEERING—CHAPTERS 1 AND 2

Chapter 1, "Introductory Concepts," introduces the essentials of chemical engineering along with unit system. Chapter 2, "Basic Principles and Introduction to Calculations," reviews well-known physical and chemical laws in continuation to Chapter 1. In addition, the principles underlying chemical engineering problems are presented.

SECTION II: FUNDAMENTALS AND PROBLEM-SOLVING PROFILE—CHAPTERS 3 AND 4

This section deals with the methodology of how to devise and evaluate numerical techniques for employing computers, through standard algorithm, in order to solve chemical engineering problems. The focus of numerical methods as presented is to translate chemical engineering problems into algorithms and implement them in a spreadsheet or programming language (e.g., MATLAB®).

Chapter 3, "Numerical Methods and Chemical Engineering Computations," is devoted to introducing numerical methods with emphasis on elementary ones. Chapter 4, "The Approach to Solve Problems by Computers," illustrates how to tackle a solution of a problem by model building.

SECTION III: BACKBONE MATERIALS OF CHEMICAL ENGINEERING—CHAPTERS 5 THROUGH 10

Section III extensively covers the basic principles underlying chemical engineering in all major processes, focusing on their scientific realization and applications. Solving problems arising in fluid mechanics, distillation, chemical reaction engineering, and others is heavily emphasized in this section. The contents of the chapters are as follows:

- Chapter 5—"Fluid Flow and Transport of Fluids"
- Chapter 6—"Heat Transmission"
- Chapter 7—"Two-Phase Diffusional Operations: Distillation and Absorption"
- Chapter 8—"Reaction Kinetics, Reactor Design, and Thermodynamics"
- Chapter 9—"Process Economics and Chemical Plant Design"
- Chapter 10—"Case Studies"

The material in each chapter in Section III, excluding Chapter 10, encompasses two sections: one covers the theoretical principles in a condensed format, and the other presents a set of numerically solved problems using interactive numerical software packages.

Chapter 10, on the other hand, introduces a number of case studies that handle practical technical problems. Many of these case studies are compiled by the author in the field of applied research, in both academy and industry.

This primer in chemical engineering would serve as a text for students majoring in chemical engineering, applied chemistry, and biochemical process engineering.

This book is designed to be used by students in a variety of ways, dependent on the level of enrollment. Possible recommendations are as follows:

- Level one, sophomore students studying a course on the basic principles in chemical engineering with programming and computer applications: Chapters 1 through 4.
- Level two, junior students seeking courses on numerical solution of chemical engineering problems or/and unit operations on fluid flow and heat transfer: Chapters 1 through 6.
- Level three, senior students: The text in Chapters 7 through 9 is recommended for a course as a continuation for unit operations to cover distillation and absorption. Kinetics and reactor design and plant design are other topics to be considered for additional course(s).

This book takes a highly pragmatic approach to presenting the principles and applications of chemical engineering and offering a valuable, easily accessible guide to solve problems using computers. MATLAB and spreadsheeting are applied in solving many examples.

Problems assigned for each chapter and found in Appendix A, ranging from simple to difficult, allow readers to gradually build their skills and be able to tackle a broad range of problems in chemical engineering. In addition, the numerous real examples throughout this book (more than 70) include computer or hand solutions, or in many cases both. Many solved example problems reinforce the concepts covered. The text is written with a view that solving problems is an essential part of the learning process. It gives a taste of the kinds of problems with which chemical engineers grapple. It puts a wealth of solved practical problems at your fingertips.

One of the main unique features of this book is that it is a convenient source in chemical engineering that can be used at ease, without the need for many comprehensive texts. It is an integrated form of a *single source*.

In conclusion, one can say that two things are important: "Science…means unresting endeavor and continually progressing development toward an aim which the poetic intuition may apprehend, but the intellect can never fully grasp." Chemical engineering is the discipline that uses a molecular understanding of matter to produce—from raw materials—all the synthetic materials humankind is using.

MATLAB® is a registered trademark of The MathWorks, Inc. For product information, please contact:

The MathWorks, Inc.
3 Apple Hill Drive
Natick, MA 01760-2098 USA
Tel: 508-647-7000
Fax: 508-647-7001
E-mail: info@mathworks.com
Web: www.mathworks.com

Acknowledgments

When authors like me get involved in writing a book, they are rather being optimistic.

Optimism is the essential ingredient for success. Getting help and support from many individuals provides inspiration that fuels the author's journey through the writing process.

In this respect, I gratefully acknowledge my indebtedness to a number of colleagues who enriched this book by their valuable contributions, either in reviewing and shaping some chapters or providing sets of solved examples and problems. Helpful suggestions offered by many of them are acknowledged. On the top of the list are Dr. Adnan Zahed (KAU, Jeddah, Saudi Arabia), for his contribution to writing Chapter 6; Dr. Mazen Al-Shalbi (KFUPM, Dhahran, Saudi Arabia), for his contribution to writing Chapter 8; and Dr. Maha Abdel-Kereem and Dr. Khaled Zohdi (both with HTI, Tenth of Ramadan, and Egypt).

In addition, help and good support were received—at an early stage of writing—from a number of colleagues: Dr. Mohamed Basyouni and Dr. Sherene Abdelhamid (both with HTI), Dr. Mohmed Helmy and Asmaa Harraz (both with Alexandria University), and Eman Abdel-Aal (Dubai Women's College).

I am especially indebted to a group of my students, at KFUPM and HTI, for including some of their homework-solved-problems in the text.

Last but not least, a word of appreciation to the staff at Taylor & Francis Group for their cooperative support and dedication in getting this book in the hands of the readers. This includes Allison Shatkin, Kari Budyk, and Robert Sims.

Author

Professor Hussein K. Abdel-Aal is emeritus professor of chemical engineering and petroleum refining, NRC, Cairo, Egypt. He received his BS in chemical engineering in 1956 from Alexandria University and MS and PhD in chemical engineering in 1962 and 1965, respectively, from Texas A&M University, College Station, Texas.

Professor Abdel-Aal joined the Department of Chemical Engineering at KFUPM, Dhahran, Saudi Arabia (1971–1985). He was the head of department for the period 1972–1974. He was also visiting professor with the Chemical Engineering Department at Texas A&M (1980–1981). In 1985–1988, he assumed the responsibility of the head of the Solar Energy Department, NRC, Cairo, before rejoining KFUPM for the period 1988–1998.

Professor Abdel-Aal conducted and coordinated projects involving a wide range of process development, feasibility studies, industrial research problems, and continuing education programs for many organizations, including Suez Oil Processing Company, Petromin in Riyadh (Saudi Arabia), Arab Petroleum Investment Corp in Dhahran (Saudi Arabia), Hagler-Baily & Company, Washington DC, Mobil/Esso Oil Companies in Libya, and Kuwait oil companies.

Professor Abdel-Aal has contributed to more than 90 technical papers, is the editor of *Petroleum Economics & Engineering*, third edition (2014), and is the main author of the textbook *Petroleum and Gas Field Processing*, second edition, (2016). Both books were published by Taylor & Francis Group/CRC Press.

Professor Abdel-Aal is listed in *Who's Who in the World* (1982) and is a member of AICHE, Sigma Si, Phi Lambda Upsilon. He is a fellow and founding member of the board of directors of the International Association of Hydrogen Energy.

Introduction: Initial Thoughts

I.1 OBJECTIVES

For a student, taking the first course in their major is an exciting but scary step into the unknown. Effectively introducing chemical engineering education to undergraduate students is usually difficult, especially when teaching introductory courses. A prime pitfall of the larger texts is that students are disinclined to read several pages of text to solve a problem.

Experience in teaching the first course has shown that there is a need for such a handy, collective, and summarized source of basic principles in chemical engineering, especially when solving problems. The methodology of solving problems and their solutions should be presented hand in hand with the fundamentals, as discussed in this text.

This primer is to be considered a survival companion for students of chemical engineering. As you navigate through the text, you can maneuver smoothly from chapter to chapter. The two main themes embodied in this text are as follows:

- *Theme No. 1*: To provide students with some guide steps on how to identify the type of problems to solve and how to target a solution. Focus will be on the use of systematic algorithms that employ numerical methods to solve different chemical engineering problems by describing and transforming the information.
- *Theme No. 2*: To present the basic principles and techniques in chemical engineering and the underlying unit operations and chemical processes in a concise and nontraditional format with numerous applications.

Let us first acquaint our readers with the field of chemical engineering. All engineers employ mathematics, physics, and the engineering art to overcome technical problems safely and economically. Yet it is the chemical engineer alone that draws upon the vast and powerful science of chemistry to solve a wide range of problems. Chemical engineers use chemistry and engineering to turn raw materials into usable products, such as petrochemicals, plastics, and pharmaceutical ingredients. They apply physical science together with mathematics and computers to processes that involve physical changes and/or chemical reactions in order to produce more valuable products.

Every scientific discipline, including chemical engineering, relies on the use of computers and software. The capabilities provided by computers, such as fast calculations, large storage capacity, and software, permit engineers to solve different problems.

Therefore, scientists depend on computational methodology in solving their problems. It is based on the following fact:

> Computational science is a function of two components: theoretical and practical one.

The *theoretical* component involves the use of *systematic algorithms* to solve different scientific problems by describing and transforming the information, while the *practical* component involves the implementation of the *computational hardware and software*.

Traditionally, in solving scientific problems, researchers use either of the following:

- Theoretical approach
- Laboratory experiment

In the theoretical approach, scientists rely upon the use of some classified *models*, in accordance with how they are derived. In general, three models are well known:

1. Theoretical models (*transparent models*): These are developed using the principles of chemistry and physics.
2. Empirical models (*black box models*): These are obtained from a mathematical or statistical process of operating data—*process identification*.
3. Semiempirical models (*gray box models*): These models represent a compromise between (1) and (2), with one or more parameters to be evaluated from plant data.

Computational science, on the other hand, is different from these two traditional approaches, the theoretical and the practical, as illustrated in Figure I.1. The scientific

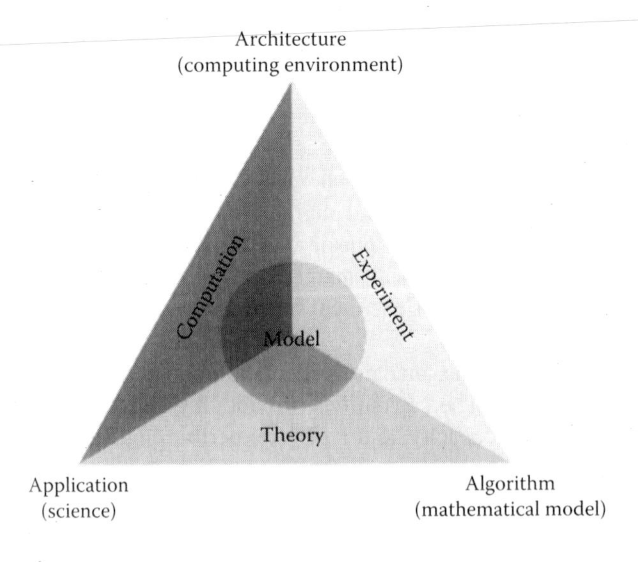

FIGURE I.1 Model of relationship among theory, computation, and experiment.

computing avenue is to gain understanding, mainly through *the analysis of mathematical models implemented on computers.*

Computers play a vital role in almost every aspect of modern chemical engineering, which may include the following applications:

1. Supercomputers to simulate complex chemical processes and plant operations
2. Computational fluid dynamics to simulate flow systems and heat transfer problems
3. MATLAB®, polymath, spreadsheeting, and others to solve many problems including differential equations
4. Aspen computer technology to simulate unit operations and processes

I.2 PROPOSED APPROACH

It was observed that frequently students had to supplement their technical knowledge when solving a problem. For example, in solving problems on flash distillation, a refresher to some basic concepts is necessary. Definitions of bubble point, dew point, and derivation of the flash equation are needed in problem solving. This book, a self-contained source on chemical engineering, will come in handy in such situations.

As an instructor in your first course, an attractive way to introduce your students to the field of chemical engineering is to present to them the following simple illustrative example of a petroleum fractionation process.

I.2.1 Illustrative Example of Petroleum Fractionation Process

Crude oil is to be fractionated into straight-run products such as gasoline and gas oil. It is heated first by heat exchangers and then desalted. Its temperature is raised next using fire heaters, before it is introduced to the fractionation tower. This process is illustrated in Figure I.2.

The next step is to explain to them how and why the crude oil is separated into products.

Well, in the transformation of raw materials (crude oil) and in the presence of energy (heat) to produce finished products, *three modes of transfer* are encountered.

Now, it may prove beneficial to examine in some detail the *transport phenomena* involved. They are known as

1. Momentum transfer (fluid flow), using a *pump*
2. Heat transfer of oil, using heat exchangers and a *furnace*
3. Mass transfer in the *distillation column* that leads to the separation of crude oil into different cuts (transfer is due to the molecular diffusion of components that separates the light from heavy)

The *physical operations* (known as unit operations and shown in the earlier example) are fluid flow, heat transfer, and distillation. They are basically based on these

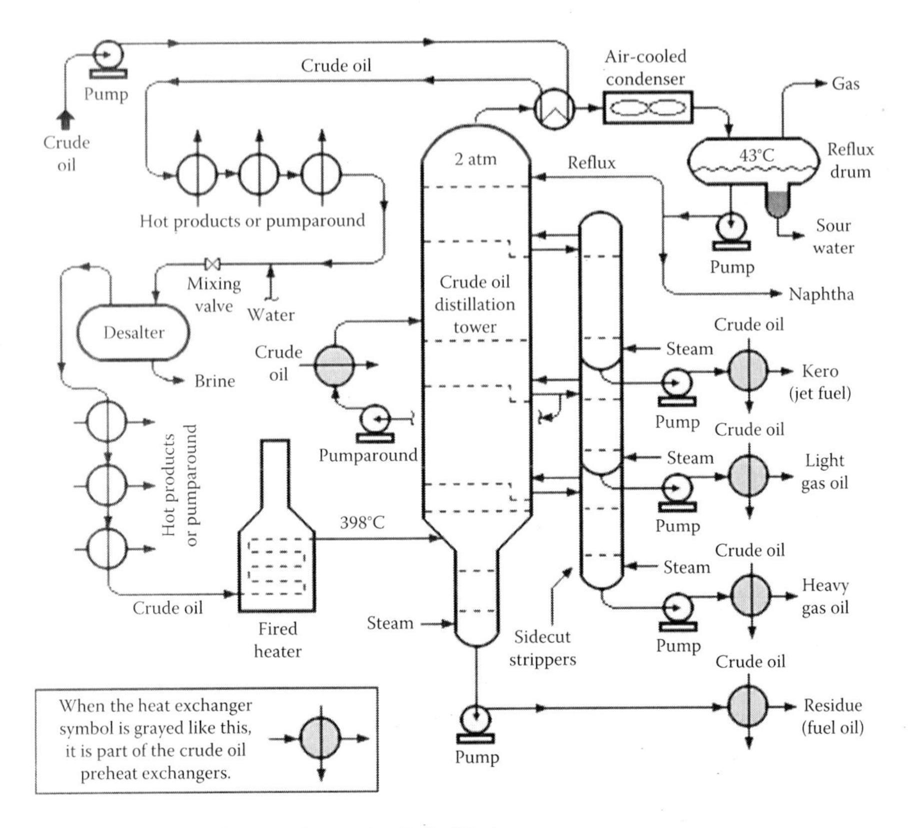

FIGURE I.2 Flow diagram for crude oil distillation.

modes of transfer. *Unit operations deal chiefly with the transfer of energy and the transfer, separation, and conditioning of materials by physical means.*

At this stage, two basic questions arise:

1. What is the mechanism(s) underlying this process?
2. How and where does it take place?

The answer to the first question deals with theory of transfer or transport, as explained, within the boundaries of our system. In answer to the second question, the combined effect of momentum, heat, and mass (MHM) is responsible for the physical changes that take place in the distillation column to produce the finished products.

Further, assume that the gasoline exiting the distillation column is introduced into what is known as a *reforming unit* in order to obtain a higher-grade gasoline. This reforming process represents a typical example of a *chemical conversion* or *chemical reaction*—known as *unit process*—where hydrocarbons undergo molecular changes and rearrangement leading to *high-octane* gasoline. *Unit processes involve primarily the conversion of materials by means of chemical reactions.* Again, it should be pointed out that the three modes of transfer, MHM, take place for operations involving chemical reactions, or chemical changes, as well.

Now, having presented this example, one should elaborate the concepts of unit operations, unit processes, and transport phenomena. This will pave the way to the students when they come to the heart of the materials embodied in the text.

I.3 ROLE OF UNIT OPERATIONS, UNIT PROCESSES, AND TRANSPORT PHENOMENA

I.3.1 Background

In early days, chemical engineering started as an applied or *industrial chemistry*. As such, it was based on the study of definitive processes. Hence, the *unit process* approach was used. Later on, it became apparent to the pioneers that certain aspects in the industrial processes were common, such as fluid flow, heat transfer, mixing, and separation technology.

This new perception led to the development and the creation of the new approach of *unit operations*, replacing the *unit process–based curriculum*.

Although the unit operations were based on first principles, they represented a *semiempirical* approach to the subject areas covered.

Later, a series of events resulted in another revolutionary response: the concept of *transport phenomena*. The subject of transport phenomena includes three closely related topics: fluid dynamics, heat transfer, and mass transfer. Fluid dynamics involves the transport of momentum, heat transfer deals with the transport of energy, and mass transfer is concerned with the transport of mass of various chemical species. These modes of transfer frequently occur simultaneously in industrial applications like distillation. The basic equations that describe the three modes of transport phenomena are also closely related and can often be analyzed by analogy. This is what is known as the *engineering science* approach to chemical engineering. The focal point of this approach is the rigorous mathematical description of all physical rate processes of mass, heat, or momentum crossing phase boundaries.

It was felt that this very last concept set the chemical engineering profession back a few decades. Graduating chemical engineers were considered to be more applied physicists than traditional chemical engineers, in terms of training.

Later, there was a return to the traditional approach in chemical engineering, primarily as a result of the Accreditation Board for Engineering and Technology.

I.3.2 Analysis and Discussion

A unit operation, as explained before, deals chiefly with the transfer of energy and the transfer, separation, and conditioning of materials by physical means. It represents a structure of logic used in the synthesis and analysis of processing schemes in the chemical engineering field (chemical and process industries). The basic underlying concept in these processes is as follows:

> All processing schemes can be composed from and decomposed into a series of individual or unit steps.

The following classification is normally adopted:

1. If a unit step involves a physical change, it is called *unit operation.*
2. If a unit step involves a chemical change, it is called *unit process.*

These unit operations cut across widely different processing applications, including the manufacture of chemicals, fuels, pharmaceuticals, pulp and paper, and processed foods. The unit operation approach serves as a very powerful form of morphological analysis, which systematizes process design and greatly reduces both the number of concepts that must be taught and the number of possibilities that should be considered in synthesizing a particular process, especially in plant design.

Most unit operations are based mechanistically upon the fundamental transport processes of mass transfer, heat transfer, and fluid flow (momentum transfer), as shown in the following table:

Type of Transport Process	Unit Operations Based Upon
Mass transfer	Distillation, solvent extraction, leaching, absorption/desorption, adsorption, and others
Heat transfer	Heat exchange, condensation, evaporation, furnaces or kilns, drying, cooling towers, and others
Momentum transfer (fluid flow)	Fluid transport (such as pumping), mixing/ agitation, filtration, clarification, thickening, or sedimentation

It is important to point out that the study of transport phenomena provides a unifying and powerful basis to understand different unit operations. The subject of transport phenomena includes three closely related topics as shown earlier; of interest is mass transfer, which is concerned with the transport of mass of various chemical species. They frequently occur simultaneously in industrial applications like distillation, where the transport of mass occurs by diffusion for various chemical species. The basic equations that describe the three modes of transport phenomena are also closely related and can often be analyzed by analogy:

Transported Quantity	Physical Phenomenon	Equation
Momentum	Viscosity (Newtonian fluid)	$\tau = -\nu \frac{\partial \rho \upsilon}{\partial x}$
Energy	Heat conduction (Fourier's law)	$\frac{q}{A} = -k \frac{dT}{dx}$
Mass	Molecular diffusion (Fick's law)	$J = -D \frac{\partial C}{\partial x}$

The equation for momentum transport is Newton's law of viscosity

$$\tau_{zx} = -\nu \frac{\partial \rho \upsilon_x}{\partial z}$$

where
τ_{zx} is the flux of x-directed momentum in the z direction
ν is μ/ρ, the momentum diffusivity
z is the distance of transport or diffusion
ρ is the density
μ is the viscosity

Newton's law is the simplest relationship between the flux of momentum and the velocity gradient.

The net flux of energy through a system, on the other hand, equals the conductivity, k times the rate of change of temperature with respect to position dT/dx.
where A is the surface area and q is the heat flow per unit time.

Finally, for mass transfer, the diffusion flux, **J**, from a higher concentration to a lower concentration is proportional to the gradient of the concentration, dc/dx, of the substance and the diffusivity, D, of the substance in the medium.

Based on this presentation, the following conclusions are drawn:

1. Chemical engineering is best defined as "the profession concerned with the creative application of scientific principles underlying the transport of mass, energy and momentum and the physical and chemical changes of matter." (American Institute of Chemical Engineers, 2003)
2. An economical method of organizing much of the subject matter of chemical engineering is based on using the concept of *unit operation,* as explained earlier. This is basically justified for two reasons:
   a. Many chemical processes can be broken down into a series of steps or operations (units), which are repeated in other processes.
   b. These individual steps or operations are based on the same scientific principles and guidelines.

However, it is believed that there is no real conflict between the two approaches: the *unit operations* and the *transport phenomena*. Most important is the development of the skills that lead to the creative engineering use of the fundamentals or a combination of both, as demonstrated and presented in our book.

REFERENCE

American Institute of Chemical Engineers. *AIChE Constitution*. http://www.aiche.org/About/WhoWeAre/Governance/Constitution.aspx, 2003. Retrieved August 13, 2011.

Section I

An Insight into Chemical Engineering

Much of the background needed for an understanding of the materials embodied in the text is basic science and engineering. To analyze and follow-up quantitatively and to optimize the underlying processes in chemical engineering a profound knowledge of theoretical background is needed. Section I, consisting of Chapters 1 and 2, highlights a *technical* introduction to the basic principles and calculation techniques introduced in the text.

Solved examples given in the chapters are straight forward direct applications. They are hand-solved. It is meant to be a kind of *refresher* for a first course in chemical engineering curriculum. Some are worked out by Excel. Application problems for Section I are included in Appendix A.

1 Introductory Concepts

As indicated by the title of this chapter, the essentials of some chemical engineering aspects are introduced. Most of the common systems of units used in calculations are presented first. Temperature (T) and pressure (P) scales are discussed in detail next, since both T and P are key process variables. For example, to design a distillation column, information is needed about the boiling temperatures of the components that should be separated. Physical gas laws that are frequently used in the calculations are described as well.

1.1 SYSTEM OF UNITS*

Our basic concepts of measurements are the following:

- Length L
- Time θ
- Mass M
- Temperature T

To express any physical property, you need *two items: a value and a unit*. For example, we say the room temperature is 25°C. Another example is the dimensions of a distillation column: 30 ft in length and 4.5 ft in diameter.

The most common systems of units are as given next:

	Length	Mass	Time	Energy
cgs	cm	g	s	J or cal
fps	ft	pound (lb)	s	ft · pdl
SI[a]	m	kg	s	Joule
Am · Eng[b]	ft	lb · mass	s or h	Btu or hp · h

[a] SI, System International.
[b] Am · Eng, American Engineering.

In addition, we have what we call *derived* units, for example, velocity = ft/s = L/θ.

The *International System of Units* (abbreviated *SI* from the French *Le Système international d'unités*) is the modern form of the metric system. The standards, published in 1960, are based on the meter–kilogram–second system rather than the centimeter–gram–second (cgs) system, which, in turn, had several variants. It comprises a coherent system of units of measurement built around seven base units, as shown in Figure 1.1.

The system has been nearly globally adopted. In the United States, metric units are not commonly used outside of science, medicine, and the government. However,

* This section is modified after International Bureau of Weights and Measures (n.d.; 2016).

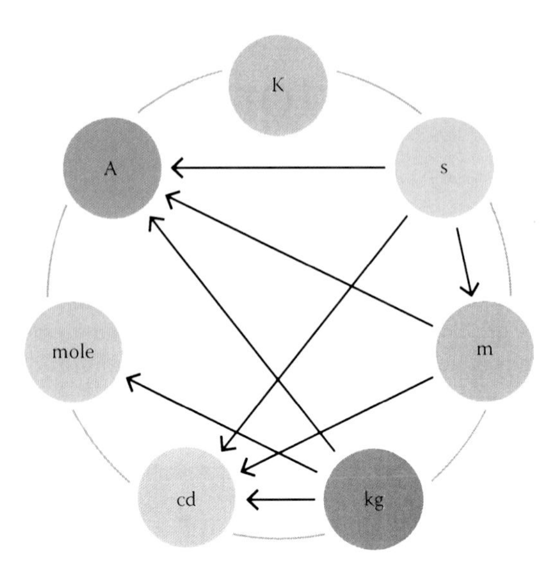

FIGURE 1.1 The seven base units of International System of Units: *Kelvin* (*temperature*), *second* (*time*), *meter* (*length*), *kilogram* (*mass*), *candela* (*luminous intensity*), *mole* (*amount of substance*), and *ampere* (*electric current*). (With kind permission from Springer: *Units of Measurement: Past, Present and Future, International System of Units*, 2009, 16, Gupta, S.V.)

U.S. customary units are officially defined in terms of SI units. The United Kingdom has officially adopted a partial metrication policy, with no intention of replacing imperial units entirely.

The basic laws that we use can be easily memorized using the following mathematical relationships:

1. Mass × velocity ⇒ *momentum* ⇒ kg · m/s
2. Rate of change of momentum $\Rightarrow d/d\theta\left[\text{momentum}\right] \Rightarrow d/d\theta[(\text{mass})\cdot(\text{velocity})]$

$$\Rightarrow (\text{mass})\cdot(\text{acceleration}) \Rightarrow \text{kg}\cdot\text{m/s}^2 \Rightarrow \textit{force} \Rightarrow \text{N}$$

3. Force × distance ⇒ *work*, *energy*, *quantity of heat* ⇒ kg · m^2/s^2 ⇒ N · m

4. Rate of change of energy ⇒ $d/d\theta$ [kg · m^2/s^2] ⇒ kg · m^2/s^3 ⇒ *power* ⇒ J/s

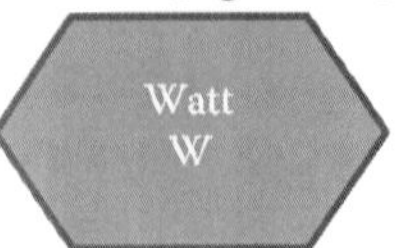

It is worth mentioning that many of the derived units in the SI system are given special names with corresponding symbols. For example, *force*, which has the units $kg \cdot m/s^2$, has been named *Newton* and given the symbol *N*. Similarly, *energy* is defined as *Newton* * *meter* or $N \cdot m$, having the units $m^2 \cdot kg/s^2$. Hence, it is named *Joule* and given the symbol *J*, and the corresponding unit of *power* is joule per second named *Watt*.

Now, let us examine the systems of units when it comes to using the conversion factor, known as g_c. Basically, we start with the Newton's law:

$$F = C \cdot m \cdot a \tag{1.1}$$

where

F is the force
m is the mass
a is the acceleration
C is a constant, whose numerical value and units depend on those selected for the units F, m, and a

- *The cgs system*: The unit of force is *dyne* where 1 g is accelerated at 1 cm/s^2. Equation 1.1 is rewritten as

$$F = [C](1\ g)(1\ cm/s^2) \Rightarrow dyne \tag{1.2}$$

  Therefore, selecting C = 1 $dyne/(g)(cm)/s^2$ and substituting this in Equation 1.2 results in F = 1 dyne.
- *The SI system*: The unit of force is *Newton* (*N*), where 1 kg is accelerated at 1 m/s^2. Equation 1.1 is rewritten as

$$F = [C](1\ kg)(1\ m/s^2) = N \tag{1.3}$$

  Therefore, selecting C = 1 $N/(kg)(m)/s^2$ and substituting this in Equation 1.3 results in F = 1 N.
- *The American engineering system*: The unit force is 1 lb_f, where 1 lb_m is accelerated at g ft/s^2, where g is the acceleration of gravity. Equation 1.1 is rewritten as

$$F = [C](1\ lb_m)(g\ ft/s^2) = 1\ lb_f \tag{1.4}$$

For Equation 1.4 to hold, the units of C have to be $C \Rightarrow (lb_f)/[(lb_m)(ft/s^2)]$.

A numerical value 1/32.174 has been chosen for the constant. The inverse of C has been given the special symbol g_c:

$$g_c = 32.174\ [(ft)(lb_m)]/[(s^2)(lb_f)] \tag{1.5}$$

1.2 PROCESS VARIABLES

Process variable is a dynamic feature of a given process. Accurate measurement of process variables is important all the way through the chemical process industry. There are four commonly measured variables that affect the performance of a process:

1. Temperature
2. Pressure
3. Flow
4. Level

We will limit our talk to the first two process variable, T and P.

1.2.1 Temperature Scales

There are three temperature scales in use today (BIPM, n.d.; International Bureau of Weights and Measures, n.d., 2016):

1. Fahrenheit
2. Celsius
3. Kelvin

Fahrenheit temperature scale is a scale based on 32 for the freezing point of water and 212 for the boiling point of water, the interval between the two being divided into 180 parts.

Celsius temperature scale, also called centigrade temperature scale, is the scale based on 0 for the freezing point of water and 100 for the boiling point of water. The following formula can be used to convert a temperature from its representation on the Fahrenheit (F) scale to the Celsius (C) value: $C = 5/9(F - 32)$. The scale is illustrated in Figure 1.2.

Kelvin temperature scale is the base unit of thermodynamic temperature measurement in the SI of measurement. Such a scale has as its zero point (absolute zero), the theoretical temperature at which the molecules of a substance have the lowest energy, as shown in Figure 1.3. Many physical laws and formulas can be expressed more simply when an absolute temperature scale is used; accordingly, the Kelvin scale has been adopted as the international standard for scientific temperature measurement. The Kelvin scale is related to the Celsius scale. The difference between the freezing and boiling points of water is 100° in each so that the Kelvin has the same magnitude as the degree Celsius.

The following equation is used for temperature *intervals* rather than specific temperatures: $1\ K = 1°C = 9/5°F = 9/5°R$.

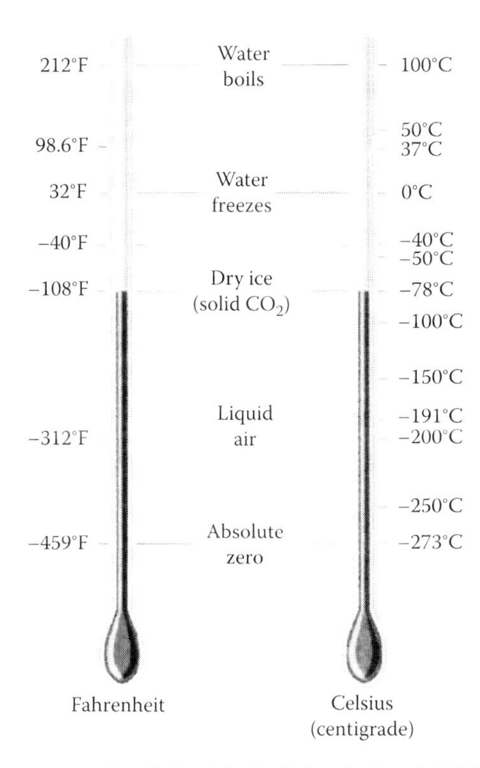

FIGURE 1.2 Temperature-scale relationship for Fahrenheit and Celsius.

1.2.2 Pressure Scale

Pascal (N/m^2) is the unit of pressure. Pressure is usually expressed with reference to either absolute zero pressure (a complete vacuum) or local atmospheric pressure. As shown in Figure 1.4, the following definitions apply:

- A perfect vacuum would correspond to absolute zero pressure.
- All values of absolute pressure are positive.
- Gauge pressures are positive if they are above atmospheric pressure.
- Gauge pressures are negative if they are below atmospheric pressure.

$$P_{abs} = P_{atm} + P_{gau} \tag{1.6}$$

As expressed in various units, *the standard atmosphere* is equal to the following:

- 1.0 atm
- 14.7 psia
- 760 mm Hg
- 29.92 in. Hg
- 33.91 ft water
- 100×10^3 Pa (pascals) or N (Newton) per sq. meter (N/m^2) = 100 kPa

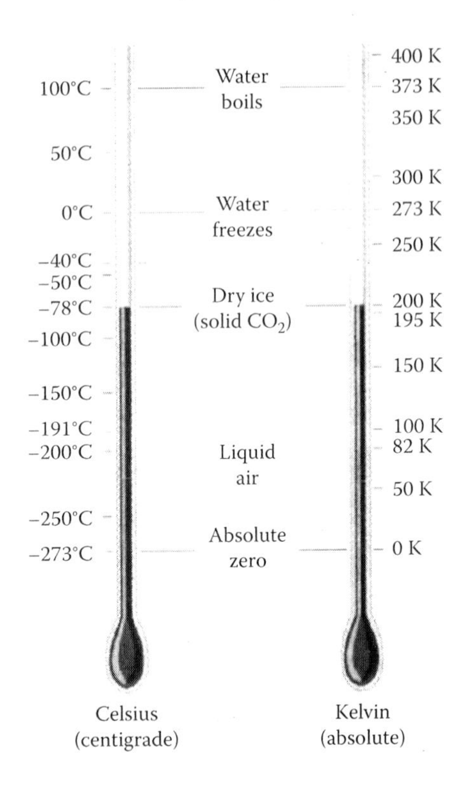

FIGURE 1.3 Temperature-scale relationship for Celsius and Kelvin.

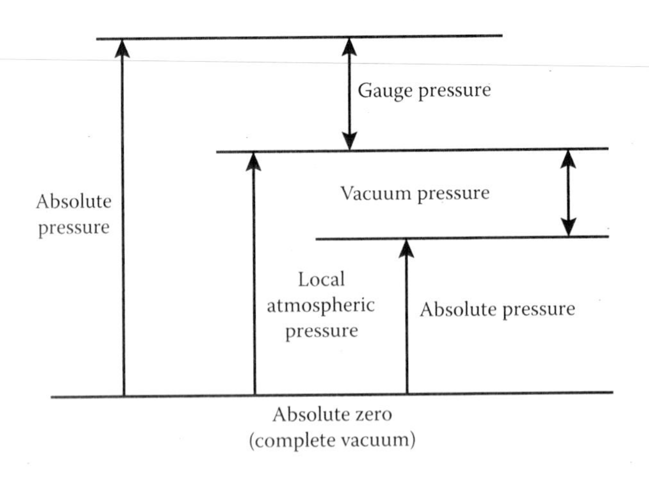

FIGURE 1.4 Illustrations of pressure relationships.

If pressure is measured by means of height of the column of liquid, other than mercury, the following equation is used to relate height to pressure:

$$P_a = P_b + \rho g h$$

where
P_a is the pressure at the bottom of the column
P_b is the pressure at the top of the column
$\rho g h$ are the density of fluid, acceleration of gravity, and height of fluid column, respectively

1.3 GAS LAWS

1.3.1 Boyle's Law

In a mathematical equation, Boyle's law is expressed as $P_1V_1 = P_2V_2 = PV = K$.

Hence, $V = K/P$.

This equation states that the volume of a given mass of gas is inversely proportional to its pressure, if the temperature remains constant. In other words, if a container, with a fixed number of molecules inside, is reduced in volume, more molecules will hit the sides of the container per unit time, causing a greater pressure.

1.3.2 Charles' Law

For an ideal gas at constant pressure, the volume is directly proportional to its temperature:

$$\frac{V_1}{V_2} = \frac{T_1}{T_2} \tag{1.7}$$

1.3.3 Dalton's Law

Mathematically, the pressure of a mixture of gases can be defined as the summation

$$P_T = \Sigma p_i = p_1 + p_2 + p_3 + \cdots + p_n \tag{1.8}$$

where p_i represents the partial pressure of each component.

A partial pressure is defined as the pressure that would be obtained if the same mass of individual gas was alone at the same V_T and at the same temperature.

Since the number of moles of a component is proportional to its partial pressure, the mole fraction of a component, A, is given by

$$X_A = \frac{p_A}{P_T} = \frac{p_A}{\Sigma p_i} \tag{1.9}$$

The most useful form in which the laws of a mixture of ideal gases can be put is as follows:

$$\text{Volume\%} = \text{Pressure\%} = \text{Mole\%}$$

1.3.4 Ideal Gas Law

From the statements given by Boyle and Charles, the following relationship is developed:

$$PV = nRT \quad (1.10)$$

where
R is the gas constant
n is the number of moles
T is the temperature
P is the pressure; both are in absolute units

In SI units, P is measured in pascals, V is measured in cubic meters, n is measured in moles, and T is measured in Kelvin (the Kelvin scale is a shifted Celsius scale, where 0.00 K = −273.15°C, the lowest possible temperature).

The gas constant (also known as the molar, universal, or ideal gas constant, denoted by the symbol R) is a physical constant that is featured in many fundamental equations in the physical sciences, such as the ideal gas law and the Nernst equation.

The dimensions of R are recognized through the following analysis:

$$R = \frac{PV}{nT}$$

where
P is the pressure
V is the volume
n is the chemical amount
T is the temperature

As pressure is defined as force per unit area, the gas equation can also be written as

$$R = \frac{\text{Force/Area} \times \text{Volume}}{\text{Amount} \times \text{Temperature}}$$

Area and volume are $(\text{length})^2$ and $(\text{length})^3$, respectively. Therefore,

$$R = \frac{\text{Force/(Length)}^2 \times (\text{Length})^3}{\text{Amount} \times \text{Temperature}}$$

Since

$$\text{Force} \times \text{Length} = \text{Work}$$

$$R = \frac{\text{Work}}{\text{Amount} \times \text{Temperature}}$$

The physical significance of R is work per degree per mole. It may be expressed in any set of units representing work or energy (such as joules), other units representing degrees of temperature (such as degrees Celsius or Fahrenheit), and any system of units designating a mole or a similar pure number that allows an equation of macroscopic mass and fundamental particle numbers in a system, such as an ideal gas. Values of R are found in Appendix B.

How much gas is present could be specified by giving the mass instead of the chemical amount of gas. Therefore, an alternative form of the ideal gas law may be useful. The chemical amount (n) (in moles) is equal to the mass (m) (in grams) divided by the molar mass (M) (in grams per mole):

$$n = \frac{m}{M}$$

By replacing n with m/M and subsequently introducing the density $\rho = m/V$, we get

$$PV = \frac{m}{M} RT$$

$$P = \rho \frac{R}{M} T$$

The ideal gas law is the equation of state of a hypothetical ideal gas. It is a good approximation to the behavior of many gases under many conditions, although it has several limitations, as shown in Figure 1.5.

Nevertheless, this form of the ideal gas law is very useful because it links pressure, density, and temperature in a unique formula independent of the quantity of the considered gas. The behavior of real gases usually agrees with the predictions of the ideal gas equation to within ±5% at normal temperatures and pressures. At low temperatures or high pressures, real gases deviate significantly from ideal gas behavior.

One of the most useful equations to predict the behavior of real gases was developed by Johannes van der Waals (1837–1923). The complete *van der Waals equation* is therefore written as follows:

$$\left\{P + \left[\frac{n^2 a}{V^2}\right]\right\}\{V - nb\} = nRT \tag{1.11}$$

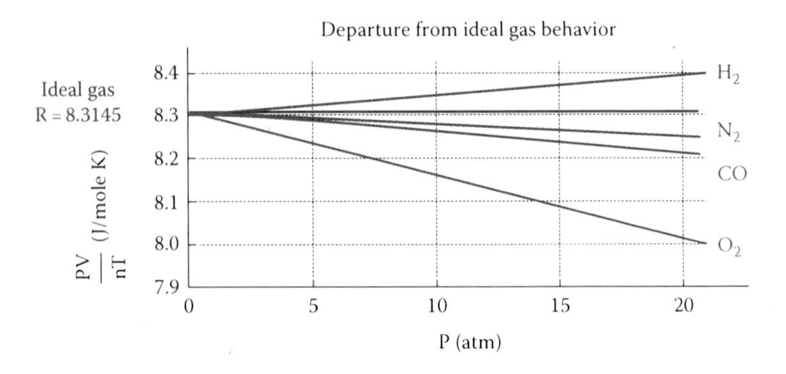

FIGURE 1.5 Deviations of gases from the ideal gas law.

From the definition of partial pressure as given before, one can say for a number of components of a gas mixture:

$$p_1V_1 = n_1RT_1 \quad \text{and} \quad p_2V_2 = n_2RT_2$$

Since

$$V_1 = V_2 \quad \text{and} \quad T_1 = T_2; \quad \frac{p_1}{p_2} = \frac{n_1}{n_2}$$

Similarly, $p_1V_1/P_TV_T = n_1RT_1/n_2RT_T$; therefore, $p_1/P_T = n_1/n_T$ or in its general form:

$$\frac{p_i}{P_T} = \frac{n_i}{n_T}, \quad \text{or} \quad p_i = P_T y_i \tag{1.12}$$

where y_i is the mole fraction of the ith component in the total mixture of n components. The next relationship provides a way to determine the volume-based concentration of any individual gaseous component

$$p_i = P_T \frac{C_i}{10^6} \tag{1.13}$$

where C_i is the concentration of the ith component expressed in ppm.

1.3.5 Raoult's Law

Raoult's law states that the vapor pressure of an ideal solution is directly dependent on the vapor pressure of each chemical component and the mole fraction of the component present in the solution. Therefore, the individual vapor pressure for each component is given by the next equation:

$$p_i = P_i^* x_i \tag{1.14}$$

where
p_i is the partial pressure of the component i in the mixture (in the solution)
P_i^* is the vapor pressure of the pure component i
x_i is the mole fraction of the component i in the mixture (in the solution)

Equating Equation 1.12 to Equation 1.14, we obtain

$$P^* x_i = P_T y_i$$

or

$$\frac{P^*}{P_T} = \frac{Y_i}{x_i} = K_i \text{ (applied only for ideal mixtures)}$$

Therefore,

$$y_i = K_i x_i \tag{1.15}$$

where K_i = f(P_T, T, and composition of mixture). If, however, K_i is independent of composition and depends only upon T, then we get the equation known as Henry's law:

$$x_i = H_i p \tag{1.16}$$

where
x_i is the solubility of a gas at a fixed temperature in a particular solvent (in units of M or mL gas/L)
H_i is Henry's law constant (often in units of M/atm)
p is the partial pressure of the gas (often in units of atm)

Henry's law states that at a constant temperature, the amount of a given gas that dissolves in a given type and volume of liquid is directly proportional to the partial pressure of that gas in equilibrium with that liquid.

1.3.6 Amagat's Law

The volume of a mixture of gases is equal to the sum of the partial volumes of each component gas. The partial volume of a component gas is the volume in which that component would occupy at the same T and P.

1.3.7 Avogadro's Hypothesis

- The molecular weight of a gas, in *kilograms*, occupies *22.4 m^3* at STP (°C, 760 mm Hg).
- The molecular weight of a gas, in *grams*, occupies *22.4 L* at STP.
- The molecular weight of a gas, in *pounds*, occupies *359 ft^3* at NTP (60°F, 760 mm Hg).

END-OF-CHAPTER SOLVED EXAMPLES

Example 1.1

The dimensional formula of a heat transfer coefficient, h, is

$$[h] = QL^{-2}\theta^{-1}T^{-1}$$

In an experimental work on the rate of heat transfer, a value of h = 396 Btu/(ft^2)(°F)(h) was obtained. Calculate the value of this coefficient in kcal/(m^2)(°C)(h)

Solution

From the conversion tables, 1 Btu/kcal = 0.252, 1 ft/1 m = 0.3048, and 1°F/1°C = 1/1.8.

By direct substitution using these conversions,

$$h = 396 \text{ Btu/(ft)}^2(°\text{F})(\text{h})$$

$$= (369)(0.252)/(0.3408)^2(1/1.8) \text{ kcal/(m)}^2(°\text{C})(\text{h})$$

$$= 1930 \text{ kcal/(m}^2)(°\text{C})(\text{h}).$$

Example 1.2

Find the average molecular weight of dry air whose volume-composition is O_2 (21%), N_2 (78%), and Ar (1%).

Solution

1	2	3	4
Comp	%	MWt	(2) (3)
O_2	0.21	32	6.72
N_2	0.78	28	21.84
Ar	0.01	20	0.2

Avg. MW 28.76

Example 1.3

Calculate the total pressure and the composition of the following gas mixture, given the partial pressure of each component:

$$CO_2: 75 \text{ mm Hg}; \quad CO: 50 \text{ mm Hg}; \quad N_2: 595 \text{ mm Hg}; \quad O_2: 26 \text{ mm Hg}$$

Solution

Substitution using Equation 1.8 gives the following:

$$P_T = \sum p_i = 75 + 50 + 595 + 26 = 746 \text{ mm Hg}$$

Mole fraction of CO_2 is obtained using Equation 1.9:

$$X_{CO_2} = \frac{75}{746} = 0.101$$

Similarly, calculations are carried out to find the mole fractions of CO, N_2, and O_2: 0.067, 0.797, and 0.035, respectively.

Example 1.4

Five hundred pounds of water is flowing a heat exchanger pipe at the rate of 20.0 ft/s. Calculate the kinetic energy of the flowing water in (ft)(lb_f).

Solution

$$\text{K.E.} = \frac{1}{2}mv^2 = \frac{1}{2}\left[(500\,lb_m)(20\text{ ft/s})^2\right]\left[\frac{1}{32.174\text{ (ft)}(lb_m)/(s^2)(lb_f)}\right]$$

$$\text{K.E.} = 3108\text{ (ft)}(lb_f)$$

Example 1.5

Calculate the volume of 1 mol of air at 20°C on top of a mountain (altitude 4.2 km), where the air pressure is approximately 60 kPa.

Solution

Apply Boyle's and Charles' laws as successive correction factors to the standard sea-level pressure of 101.3 kPa:

$$V_2 = (22.4\text{ L})\left(\frac{293\text{ K}}{273\text{ K}}\right)\left(\frac{101.3\text{ kPa}}{60\text{ kPa}}\right) = 41\ L$$

The standard molar volume 22.4 L/mole is applicable at standard temperature and pressure STP.

Example 1.6: To Check the Validity of the *van der Waals Equation*

Compare the experimental PV data for propane gas to the predicted values as shown next:

$P = ((R * T)/(V - b)) - (a/V^2)$

P Experimental	A	V	B	R	T	Predicted P	Delta P
8.3	8.61	5	0.0909	0.083	430	6.9	1.4
3.5	8.61	10	0.0909	0.083	430	3.5	0.0
2.5	8.61	15	0.0909	0.083	430	2.4	0.1
2.1	8.61	20	0.0909	0.083	430	1.8	0.3
1.4	8.61	25	0.0909	0.083	430	1.4	0.0
1.4	8.61	30	0.0909	0.083	430	1.2	0.2
1.2	8.61	35	0.0909	0.083	430	1.0	0.2
1.1	8.61	40	0.0909	0.083	430	0.9	0.2
0.8	8.61	45	0.0909	0.083	430	0.8	0.0
0.65	8.61	50	0.0909	0.083	430	0.7	−0.1
0.7	8.61	55	0.0909	0.083	430	0.6	0.1
0.65	8.61	60	0.0909	0.083	430	0.6	0.1

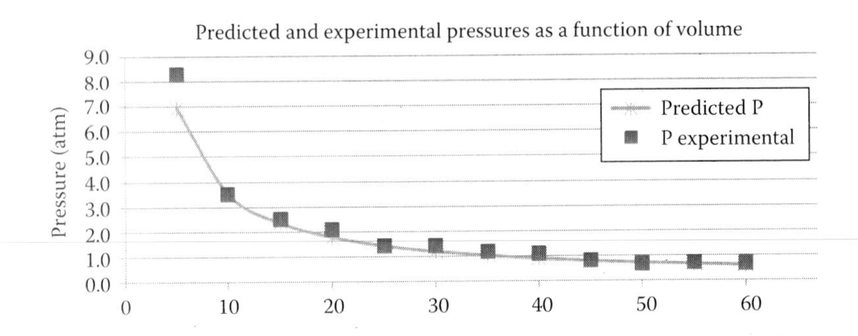

Conclusion: The PV values for both experimental and predicted ones match closely.

Example 1.7

CO is produced by the reaction:

$$CO_2\ (g) + C\ (s) \rightarrow 2CO\ (g)$$

At equilibrium conditions of 1000°C and a total pressure of 30 atm, 17.0 mole% of the gas is CO_2. If the total pressure is reduced to 20 atm, calculate the CO_2 concentration.

Solution

$p_i = P_T y_i$.
P_T is the sum of $p_i = p_{CO_2} + p_{CO}$.

First: At 30 atm,

$$K_p = \frac{p_{CO}^2}{p_{CO_2}}$$

$$= \frac{(30^2)(0.83)^2}{(30)(0.17)} = 121.5 \text{ atm.}$$

Second: At 20 atm, $K_P = 121.5 = p_{CO}^2 / p_{CO_2} = \{20 - p_{CO_2}\} / p_{CO_2}$.
Solving for p_{CO_2}, we obtain $p_{CO_2} = 2.55$ atm.
Percent CO_2 = [2.55/20] × 100 = *12.75*.

Example 1.8

Compute the work in joules if a steady force of 3 lb_f is needed to push 5 lb_m in a distance of 100 ft.

Solution

$$\text{Work (W)} = \text{Force} \times \text{distance} = 3\ lb_f \times 100\ ft = 300\ ft \cdot lb_f$$

$$= (300\ ft \cdot lb_f)(1/0.7367\ J/ft \cdot lb_f)$$

$$= 406.7\ J$$

Example 1.9

Calculate the molar volume (ft³/lb · mole) and the density (lb/ft³) of methane gas at 60°F (519.67°R) and a pressure of 3.5 atm. Molecular weight (M) for methane is 16.04.

Solution

$$\text{Molar volume} = RT/P = [(0.7302\ ft^3 \cdot atm/lb \cdot mole \cdot {}^\circ R)519.67^\circ R(1/3.5\ atm)]$$

$$= 108.4\ ft^3/lb \cdot mole$$

Methane density is

$$\rho = (PM)/(RT) = \frac{(3.5\ atm)(16.04\ (lb/lb_m)}{[(0.7302)\ (lb_m R)/(ft^3 \cdot atm)](519.67)}$$

$$= \mathit{0.148\ lb/ft^3}$$

Example 1.10

Calculate the quantity of heat required to raise the temperature of 40 g of ammonia gas from 70°C to 100°C. Specific heat of ammonia is 0.51 Btu/lb·°F.

Solution

$$\begin{aligned} Q \text{ (heat quantity)} &= \text{(mass)(specific heat)}(\Delta T) \\ &= (40/453.6)\ (\text{lb/g})(0.51)\ (\text{Btu/lb}\cdot{}^\circ\text{F})30(1.8)({}^\circ\text{F}/{}^\circ\text{C}) \\ &= 2.4286\ Btu \end{aligned}$$

Example 1.11

Find the absolute pressure (mm Hg) of a gas under draft of 5.0 in water. The barometric pressure reads 745.0 mm Hg.

Solution

$$\begin{aligned} \text{The draft reading} &= (5.0\ \text{in.})(25.4\ \text{mm/in.})\ (1/13.6\ \text{mm Hg/mm water}) \\ &= 9.338\ \text{mm Hg} \end{aligned}$$

Therefore, the absolute pressure in the duct is 745.0 + (−9.338) = 735.66 mm Hg.

REFERENCES

BIPM, n.d. SI Brochure, Section 2.1.1.5, http://wwwl.bipm.org/en/si/si_brochure/chapter2/2-1/2-1-1/kelvin.html.

International Bureau of Weights and Measures, n.d., Brief History of the SI, http://www.bipm.org/en/si/history-si/, retrieved April 21, 2009.

International Bureau of Weights and Measures," *Wikipedia*, last modified date Aug. 2, 2016, https://en.wikipedia.org/wiki/International_Bureau_of_Weights_and_Measures.

2 Basic Principles and Introduction to Calculations

In this chapter, a review of some of the well-known physical and chemical laws is given in continuation to Chapter 1. In addition, the principles underlying chemical engineering problems are presented, showing readers how to apply such principles to the different topics covered in the text.

Definitions of many of the concepts and terms that underlie much of the chemical and process industries are summarized, such as unit operations and unit processes, to name a few. Design variables for sizing piping and process equipment are briefly presented. Systems of units, different temperature scales, and pressure scales are described.

Much of the materials given in this chapter would add basic knowledge to the students and enhance the ability in problem solving.

Section I, consisting of Chapters 1 and 2, would pave the way for the readers, especially when it comes to topics on unit operations and plant design. It may be considered as a supplement to the chapters found in Section III.

2.1 THE TWO BUILDING BLOCKS OF CHEMICAL ENGINEERING: UNIT OPERATION AND UNIT PROCESSES

An economical method of organizing much of the subject matter of chemical engineering is based on using the concept of *unit operation*. This is because of two reasons:

1. In many chemical processes, each one can be broken down into a series of steps or operations that are repeated in other processes.
2. These individual steps or operations are based on the same scientific principles and guidelines.

A unit operation represents a basic physical operation in a chemical process plant. Examples are distillation, absorption, fluid flow, heat transfer operations, evaporation, extraction, drying, crystallization, and filtration. Fundamentals pertaining to a given unit operation are the same regardless of its industrial applications. This is how pioneers came up with the term *unit operation*.

Unit operations deal mainly with the transfer and the change of both materials and energy primarily by physical means, arranged as needed by a chemical process. The following is a partial list of some important unit operations:

- *Fluid flow*: It deals with the governing principles for flow and transportation of fluids.
- *Heat transfer*: It deals with the underlying principles of heat transfer by different modes.
- Diffusional mass transfer unit operations include *distillation, absorption, extraction, and drying*. Separation in these unit operations is accomplished by the transfer of molecules from one phase to the other by diffusion.

Unit process, on the other hand, involves a chemical conversion of materials in an industrial chemical plant. A good example will be the reaction of nitrogen with hydrogen to produce ammonia. A unit process represents the chemical equivalent of a unit operation, as illustrated next.

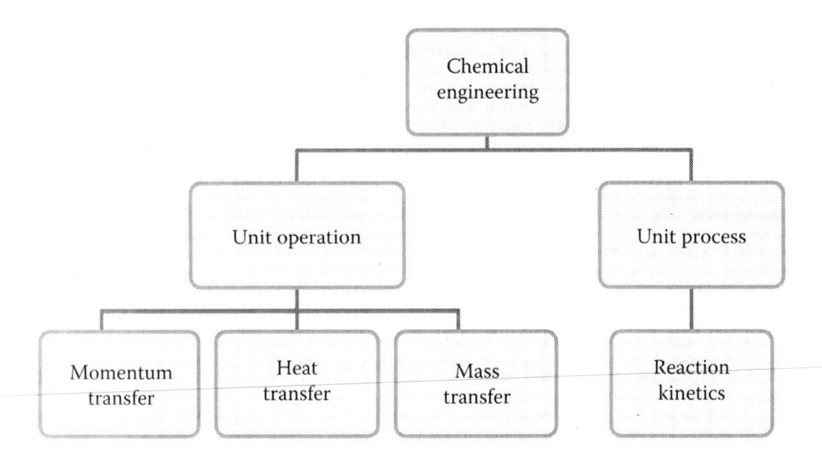

2.2 CHEMICAL PROCESS INDUSTRY

Chemical process industry (CPI) is the term that encompasses all chemical-producing industries. In the CPI, we encounter two types of processes:

1. One process basically involves *physical separation or physical changes, which is the unit operation*. Examples include crude oil refining and water desalination. Here, we have to write material balance (MB) equations to solve for the unknown parameters.
2. Another process basically involves *chemical reaction or chemical changes, which is the unit process*. This includes the manufacture of ammonia, cement, and others. Here, we have to use the stoichiometric chemical equations to calculate the yield.

Many industries would involve *a chemical reaction process* followed by a physical separation step. In addition to these distinctive chemical reaction processes or physical separation operations, other unit operations may be needed.

Physical separation operations (distill, absorption) are known as unit operations that involve mass transfer between phases. They are called diffusional unit operations (UO).

Within the field of chemical engineering, two broad groups of engineers are identified:

1. The design and the operation of plants in chemical industry and related processes. This group is known as *chemical process engineers.*
2. The development of new chemicals or products ranging from foods and beverages to cosmetics to cleaners to pharmaceutical ingredients, among many other products. This group is known as *chemical product engineers.*

2.3 CHEMICAL PLANT

A chemical plant is one that belongs to the *chemical industry*. Most processes in the chemical industry involve a *chemical change*. The term *chemical changes* should be understood to include not only *chemical reactions*, for example, $N_2 + 3H_2 \rightarrow 2NH_3$, but also *physicochemical* changes, such as the separation and purification of the components of a mixture, crude oil fractionation, and water desalination are typical examples.

In general, chemical plants perform three distinctive functions:

1. Preparation and purification of raw materials
2. Chemical reactions
3. Separation and purification of products

Purely mechanical changes are usually not considered part of the chemical processes.

A chemical plant is schematically shown in Figure 2.1.

As a result of the transformation of raw materials in the presence of energy-producing finished products, chemical processes are analyzed and studied using the following chemical engineering tools:

1. Mass and energy balances
2. Thermodynamics
3. Reaction kinetics and catalysis

FIGURE 2.1 Chemical plant with input of raw materials and energy source.

4. Unit operations
5. Instrumentation and control
6. Economics

These six subjects, known as *the backbone courses* in a typical BS degree curriculum, are closely dependent on each other, but their principles can be studied individually.

2.4 PROCESS DESIGN

It involves the design of a process for a given desired physical and/or chemical transformation of materials. Process design is the heart of plant design, and it can be considered to be the *summit* of chemical engineering, bringing together all of the components of that field.

Process design covers two distinctive cases:

1. The design of new facilities.
2. The modification or expansion of existing facilities. The design starts at a conceptual level and ultimately ends in the form of fabrication and construction plans.

A process design includes the following major items:

1. Process flow sheet
2. Mass balance on the overall process and on each unit in the process
3. Energy balance for all units, including heat exchangers requirements
4. Specs of pump capacities, flow, and pressure requirements
5. Specs of size and configuration of chemical reactors, distillation columns, absorbers, and storage tanks
6. Estimation of utility requirements, such as steam, water, electricity, compressed air, and fuel
7. Economic evaluation that includes an estimate of capital investment, operating costs, and profitability analysis

2.5 ROLE OF MATERIAL AND ENERGY BALANCES, THERMODYNAMICS, AND KINETICS

1. *Material and energy (M and E) balances*: The laws of M and E balances state that neither mass nor energy can be created or destroyed, but they may be changed in form. MB is based on the conservation of mass. Similarly, energy balance (EB) is based on the law of conservation of energy. Often energy changes in form (observe: carbonate + heat → calcium oxide + carbon dioxide).
2. *Thermodynamics*: It deals with the transformation of energy from one form to another (EB is an expression of the *first law* of thermodynamics). The *second law* states that in a process of heat transfer alone, energy may be transferred only from higher T to a lower T.

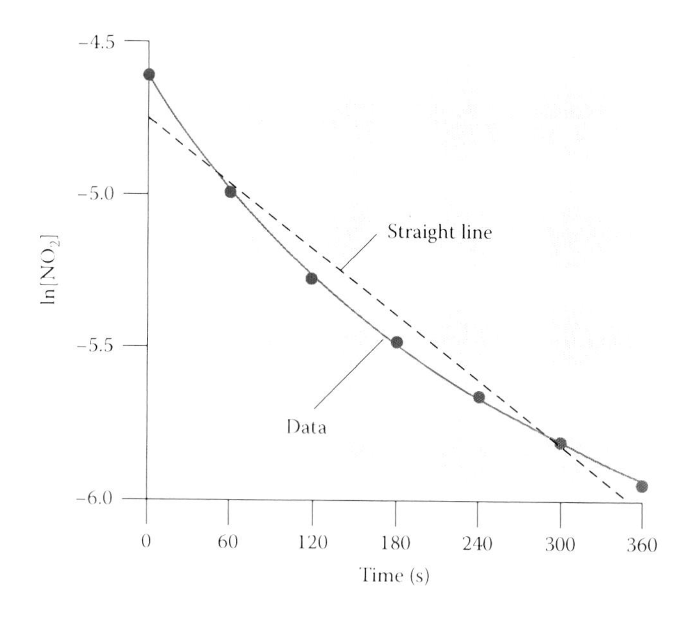

FIGURE 2.2 Thermal decomposition of NO_2.

Thermodynamics is needed in order *to judge the feasibility and efficiency of a chemical process, using the well-known relationship:*

$\Delta G = -RT \ln k$ (*the value of* ΔG *has to be negative for a reaction to be feasible*)

In other words, given the value of ΔG, calculate k (the chemical reaction equilibrium constant), hence the composition of phases at equilibrium).

3. *Kinetics*: This law is used to determine the rate at which a chemical compound reacts, that is, finding the rate of reaction, k. It is needed in the design of chemical reactors.
4. *Kinetics* versus thermodynamics.

As stated earlier, the maximum extent to which a chemical reaction can proceed may be calculated by thermodynamics, which occurs at equilibrium. It follows that the net rate of a chemical reaction must be zero at this equilibrium point. Thus, a plot of reaction rate (r) versus time would always approach zero, as the time approaches ∞, as illustrated in Figure 2.2 for the thermal decomposition of NO_2 to $NO + O_2$.

2.6 DESIGN VARIABLES FOR SIZING PIPING AND PROCESS EQUIPMENT

1. Physical unit operations include the following:
   a. *Fluid flow in pipes*: Noncompressible.
   Given: Q (ft^3/m) and v (ft/m); you get → Di and then → ΔP
   Given: P (fluid pressure) and S (tensile strength of material); you calculate → schedule no, and then determine → t (pipe thickness)

b. *Heat transfer*: $Q = mc_p\Delta t = UA\Delta T_{av}$.
c. *Distillation*: The variables are N (number of trays), D (column diameter), and R (reflux ratio).

2. Chemical reactors.
The relationship is between three variables: V (volume of reactor, ft^3), θ (residence time, s), and q (flow rate, ft^3/s).
Therefore, $V = (q)(\theta)$.

2.7 BASIC DEFINITIONS

- Equilibrium constant for a chemical reaction (reversible) is defined as

$$k = \frac{\text{Concentration of products}}{\text{Concentration of reactants}}.$$

- Distribution constant for a component in a mixture (vapor/liquid equilibrium) is defined as $K_i = y_i/x_i$.
- The three modes of transfer in chemical processes are *momentum*, *heat*, and *mass*.
- The role of ΔG in chemical reactions is illustrated by the following equation: $\Delta G = -RT \ln K$.
- A differential balance indicates what happens in a system at an instant of time; each term in the balance is a rate (*quantity/time*), and it is applied for a *continuous process*.
- An integral balance describes what happens between two instants of time. Each term in the equation is an amount of the balanced (*quantity*). It is applied to a *batch process*.

2.8 CONVERSION OF PHYSICAL EVENTS AND PRINCIPLES TO MATHEMATICAL FORMULA

As an example, flash distillation problem is shown in Figure 2.3. It is required to calculate the bubble point (BP), the dew point (DP), the flow rates of the streams leaving the flash distillation column, and their composition.

To illustrate this application, only the calculation of the BP is done, as illustrated next.

Definition: The BP is physically defined as the temperature at which the first bubble comes out as vapor. One can express this physical statement by saying that $\sum y_i = 1$, where y_i is the composition of component i in the vapor phase.

Then, we carry on one more step to say that since $y_i = K_i x_i$, we get $\sum k_i x_i = 1$. Now, the final mathematical relationship is given by

$$\sum k_i x_i - 1 = f(T_{BP}),$$

which goes to zero at the right assumed value of T.

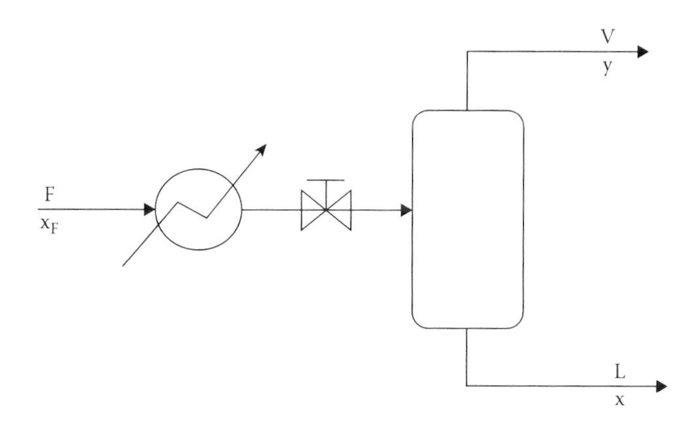

FIGURE 2.3 Flash calculation.

This equation represents the mathematical function for the physical statement of the BP. Therefore, the solution for the BP becomes a trial and error problem by solving the function, since K_i is a function of T or $K_i = f(T_i)$.

2.9 BASIC LAWS

Balance equations for momentum, mass, and energy transfer provide the broad foundation for much of the physical problems we encounter in chemical engineering curriculum. They represent important starting steps for developing phenomenological events especially in dealing with fluid mechanics (momentum transfer), heat and mass transfer, and reaction studies as well.

2.9.1 Mass and Energy Conservation: Material Balance and Heat Balance

1. The law of conservation of mass, in its most compact form, states that *matter is neither created nor destroyed.*

   In material balance calculations, it is stated that the total mass of all materials entering in a process must equal the total mass leaving, plus any materials accumulating:

   $$\text{Input} = \text{Output} + \text{Accumulation}$$

   At steady-state conditions, *Input = Output.*
2. The law of conservation of energy (called the first law of thermodynamics), in its most compact form, states that energy is neither created nor destroyed.

When the two laws are merged together, we get what is known as the *law of conservation of mass–energy: the total amount of mass and energy in the universe is constant*, where mass and energy can interconvert.

For heat balance calculations for a system, the energy or heat input in the form of enthalpy processed by materials in, plus any heat added, should be equal to heat out.

2.9.2 Thermodynamics

The four laws of thermodynamics define the fundamental physical quantities (temperature, energy, and entropy) that characterize thermodynamic systems. The laws describe how these quantities behave under various circumstances and forbid certain phenomena (such as perpetual motion):

1. *Zeroth law of thermodynamics*: If two systems are both in thermal equilibrium with a third system, then they are in thermal equilibrium with each other. This law helps to define the notion of temperature.
2. *The first law of thermodynamics*: Heat and work are forms of energy transfer. Energy is invariably conserved; however, the internal energy of a closed system may change as heat is transferred into or out of the system or work is done on or by the system. It is a convention to say that the work that is done by the system has a positive sign and connotes a transfer of energy from the system to its surroundings, while the work done on the system has a negative sign.
3. *The second law of thermodynamics*: An isolated system, if not already in its state of thermodynamic equilibrium, spontaneously evolves toward it. Thermodynamic equilibrium has the greatest entropy among the states accessible to the system. Perpetual motion machines of the second kind are thus impossible.
4. *The third law of thermodynamics*: The entropy of a system approaches a constant value as the temperature approaches zero. The entropy of a system at absolute zero is typically zero and, in all cases, is determined only by the number of different ground states it has. Specifically, the entropy of a pure crystalline substance at absolute zero temperature is zero.

2.9.3 Some General Tips

2.9.3.1 *Parts-Per* Notation (Schwartz and Warneck, 1995)

The *parts-per* notation is a unit that deals with very small traces of species within a mixture of gases or liquids. Parts-per million (ppm) and parts-per billion (ppb), as well as parts-per trillion (ppt) (American definition of trillion 1012), refer to mass or mole ratios and communicate how many parts of the species are present per million, billion, or trillion parts of the mixture. Generally, mass ratios are used when dealing with liquids and mole ratios are used when dealing with gases.

For example, let us say the air around us contains 20 ppm He (helium). This means that, if one assumes that a molar basis is being used, for every million moles of air, there are 20 moles of helium. If the example was in terms of ppb, this would mean that for every billion moles of air, there are 20 moles of He.

2.9.3.2 Frequently Asked Questions

- What is the difference between distillation and evaporation?
  Hint: Recovery of glycerin from dilute solution as a by-product in soap making.
- How do you determine if a chemical reaction is feasible to carry on?
  Hint: Calculate ΔG.
- The chemical engineer, in general, is interested *not* so much in the state of *equilibrium* as in the *rate* at which a process is taking. Explain?
  Hint: At equilibrium the driving force = zero.
- Differentiate between the concept of equilibrium in chemical and physical systems.
  Answer: For the reversible reaction, A + B <--> C, K (reaction rate) = [C]/[A] · [B], while for distillation, the distribution constant is $K_i = y_i/x_i$.
- Define the roles of kinetics and thermodynamics in chemical engineering.
  Answer: While chemical kinetics is concerned with the rate of a chemical reaction, thermodynamics determines the extent to which reactions occur, that is, to find the yield at equilibrium. In a reversible reaction, chemical equilibrium is reached when the rates of the forward and reverse reactions are equal (the principle of detailed balance) and the concentrations of the reactants and products no longer change.
- What are the main six categories of chemical reactions?
  Answer: (1) Combustion, (2) Synthesis, (3) Decomposition, (4) Single displacement, (5) Double displacement, and (6) Acid base
- Describe what is meant by the following: (a) Isothermal process, (b) Isobaric process, (c) Isometric process, (d) Adiabatic process, and (e) Isentropic process
  Answer: (a) dt = 0, constant T, (b) dP = 0, constant P, (c) dV = 0, constant V, (d) Q = 0, (e) constant entropy

END-OF-CHAPTER SOLVED EXAMPLES

Example 2.1

Sodium sulfite is used to remove dissolved oxygen from boiler-feed water to avoid corrosion, as given by the equation:

$$2\ Na_2SO_3 + O_2 \rightarrow 2Na_2SO_4$$

As a chemical engineer in charge of the utilities department, calculate how many pounds of S.S. are required theoretically, to remove dissolved O_2 from 1 million gallon of feed water that contains 10.0 ppm of O_2?

C	D	E	F	G	H
Solution					
	lb/gal		lb		
1×10^6 gallons water=	8.33	1.E+06	8.33E+06		

		lb O2	lb water	lbm/lb	lbm
Oxygen content in feed water=		10	1.00E+06	1/32	2.60

Na_2SO_3 required =2.6x [2/1]x 126 =	656 lb

Example 2.2

CO is produced by the reaction:

$$CO_2(g) + C\ (s) \rightarrow 2CO\ (g)$$

At equilibrium conditions of 1000°C and a total pressure of 30 atm, 17.0 mole% of the gas is CO_2. If the total pressure is reduced to 20 atm, calculate the CO_2 concentration.

Solution

$p_i = P_T y_i$.
P_T is the sum of $p_i = p_{CO_2} + p_{CO}$.

First: At 30 atm,

$$K_p = \frac{p_{CO}^2}{p_{CO_2}}$$

$$= \frac{(30^2)(0.83)^2}{(30)(0.17)} = 121.5\ \text{atm.}$$

Second: At 20 atm, $K_P = 121.5 = p_{CO}^2/p_{CO_2} = \{20 - p_{CO_2}\}/p_{CO_2}$.
Solving for p_{CO_2}, we obtain $p_{CO_2} = 2.55$ atm.
Percent CO_2 = [2.55/20] × 100 = 12.75.

Example 2.3

Calculate the volume of 1 mol of air at 20°C on top of a mountain (altitude 4.2 km), where the air pressure is approximately 60 kPa.

Solution

Apply Boyle's and Charles' laws as successive correction factors to the standard sea-level pressure of 101.3 kPa:

$$V_2 = (22.4\ \text{L/mole})\ \frac{293\ \text{K}}{273\ \text{K}}\ \frac{101.3\ \text{kPa}}{60\ \text{kPa}}$$

$$= 41\ \text{L}$$

The standard molar volume 22.4 L/mole is a value that it is valid only at STP.

Example 2.4

For the industrial production of CO_2, fuel oil is injected into a furnace to be burned with exactly 1.5 times the theoretical amount required for complete combustion. Calculate the exact composition of the stack gases, and determine the CO_2 purity of the product to be sold. For simplicity, assume the fuel oil to be $C_{12}H_{26}$.

Solution

$$C_{12}H_{26} + 18.5O_2 \longrightarrow 12CO_2 + 13H_2O$$

Using 100 kg·mole of fuel oil as a basis:

O_2 required = 1.5 × 18.5 × 100 = 2775 kg·mole

N_2 in with the O_2 = 10,307 kg·mole

Excess O_2 = 2775 − 1850 = 925 kg·mole

Results based on 100 kg·mole of fuel are as follows:

Products	kg·mole	%
CO_2	(100)(12) = 1,200	1,200/13,732 = 8.7
H_2O	(100)(13) = 1,300	9.5
O_2	925	6.7
N_2	10,307	75.1
Total	13,732	100

Example 2.5

It is required to concentrate an extracted orange juice using vacuum evaporator. Conditions of input and output are as indicated in the diagram.

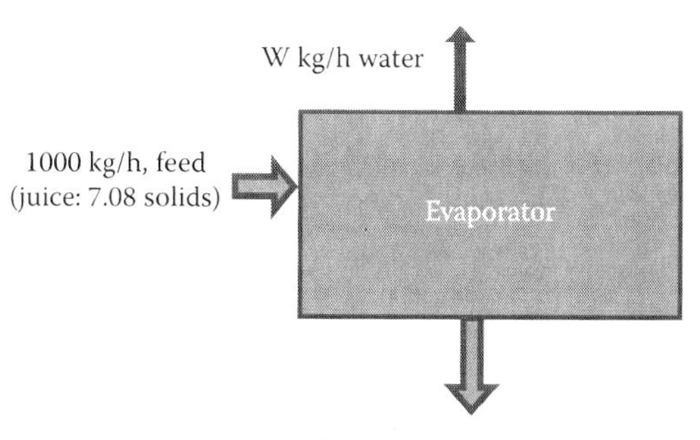

Choosing 1000 kg/h as a feed, calculate the values of W and B.

Solution

This is a physical separation process. We have two unknowns to calculate by using total MB (TMB) and component MB (CMB).

TMB:

$$1000 = W + B \tag{2.1}$$

CMB on solids:

$$1000(0.0708) = W(0.0) + B(0.58) \tag{2.2}$$

Solving the two equations in two unknowns, we obtain $W = 878\ kg/h$ and $B = 122\ kg/h$.

Example 2.6

Combustion of heptane (C_7H_{16}) is used as a source to produce *dry ice*. Calculate how many kilograms of heptane must be burned to produce 500 kg of dry ice, assuming that 50% of the CO_2 is converted into dry ice.

Solution

The chemical equation of the combustion of heptane states the following:

$$1\,\text{mole } C_7H_{16} + 11\,\text{mole } O_2 \rightarrow 7\text{ mole } CO_2 + 8\text{ mole } H_2O1.$$

Therefore, (500 kg dry ice)(1 kg CO_2/0.5 kg dry ice)(1 kg·mole CO_2/44 kg CO_2) (1 kg·mole C_7H_{16}/7 kg·mole CO_2)(100 kg C_7H_{16}/1 kg·mole C_7H_{16}) = *325 kg* C_7H_{16}.

Example 2.7

Calculate the kinetic energy of a missile moving at 12,000 mile/h above the earth, with an acceleration due to gravity of 30 ft/s².

Solution

We arbitrarily take 1 lb mass as a basis.

$$\begin{aligned}
\text{K.E.} &= \frac{1}{2}mv^2 \\
&= \frac{1}{2}(1\ \text{lb}_m)[1\ \text{lb}_f \cdot (s^2)/32.2\ \text{ft}\cdot\text{lb}_m][(12{,}000\ \text{miles})^2/h^2] \\
&\quad \times[(5280\ \text{ft})^2/\text{mile}^2][h^2/(3600\ s)^2 \\
&= 4.81\times10^6\ \ (ft)(lb_f)
\end{aligned}$$

Example 2.8

The well-known equation to calculate the pressure drop due to friction losses for fluids in pipes is given by $\Delta P = [2fL\rho v^2/D]$, where ΔP is the pressure drop, v is the velocity, L is the length of pipe, ρ is the density of fluid, and D is the diameter of pipe.

Find out if this equation is dimensionally consistent. Support your answer.

Solution

$$\Delta P = [2fL\rho v^2/D] = (ft/ft)(lb_m/ft^3)(ft/s)^2[lb_f \cdot (s^2)/lb_m \cdot ft]$$
$$= lb_f/ft^2$$

The equation is consistent as far as the dimensions are concerned; f is dimensionless.

Example 2.9

Lithium hydroxide is capable of removing CO_2 by absorption, particularly for manned spacecraft according to the following reaction:

$$2LiOH(s) + CO_2 \rightarrow LiCO_3(s) + H_2O$$

Assuming that 1 kg of CO_2 is released per day per person, calculate the amount of LiOH required.

Solution

Basis *1 kg CO_2*

$$[1\,kg\ CO_2][1\,kg \cdot mole\ CO_2/44.01\,kg\ CO_2][2\,kg \cdot mole\ LiOH/1\,kg \cdot mole\ CO_2]$$
$$[23.94\,kg\ LiOH/1\,kg \cdot mole\ LiOH] = 2.18\,kg\ LiOH$$

Answer: *2.18 kg LiOH* per day per person is needed.

Example 2.10

A solvent recovery system delivers an inert gas, saturated with benzene vapor (C_6H_6). This gas is at 70°F and 750 mm pressure. The gas is compressed to 5 atm and cooled to 70°F, after compression.

Calculate how many pounds of benzene are condensed per 1000 ft^3 of the feed gas. The vapor pressure of benzene at 70°F is 75 mm.

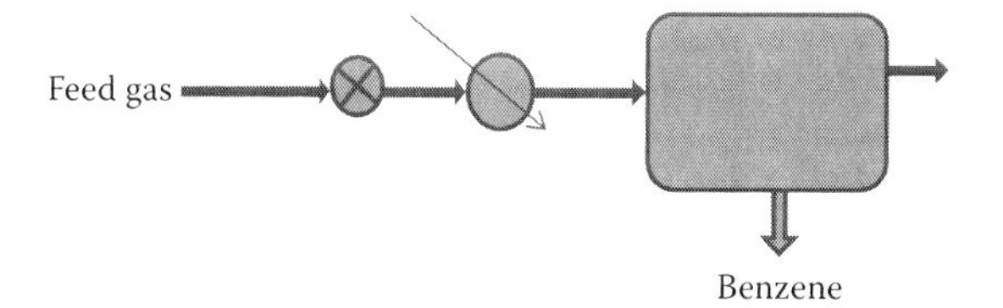

Solution

The basis is 1000 ft^3 of feed gases consisting of benzene plus inert. Applying volume % = pressure %, where the total pressure is 750 mm, while the partial pressure of benzene is 75 mm, the volume % of the inert = (750 − 75)/750.

The volume of inert gas in the feed = (1000)(750 − 75)/750 = 900 ft^3.

Then the volume of benzene = 100 ft^3.

Convert the volumes to moles: moles of inert = 2.3 and moles of benzene = 0.2556.

The ratio of benzene to inert, before compression, is 0.2556/2.3 = 0.1111.

Now after compression and cooling, a portion of benzene will condense and be separated. Ratio will be 75/[(760 × 5) − 5] = 0.0201. Hence (0.1111 − 0.0291) = 0.0910 mole of benzene is condensed per mole of inert.

For 2.3 mol of inert, benzene separated is 0.0910 × 2.3 × 78.1 = *16.35 lb*.

REFERENCE

Schwartz, S.E. and Warneck, P. Units for use in atmospheric chemistry. *Pure & Applied Chemistry* 67: 1377–1406, 1995. Retrieved on March 9, 2011, http://www.iupac.org/publications/pac/1995/pdf/6708x1377.pdf.

Section II

Fundamentals and Problem-Solving Profile

Chapter 3 Numerical Methods and Chemical Engineering Computations
Chapter 4 The Approach to Solve Problems
Section II encompasses Chapters 3 and 4. Chapter 3, "Numerical Methods and Chemical Engineering Computations," is devoted to the introduction of numerical methods, which are often divided into two main categories:

1. *Elementary ones* such as finding the root of an equation, integration of a function or solving a set of linear equations.
2. *Intensive ones* which are often needed for the solution of practical problems.

Emphasis in Chapter 3 is placed on elementary methods, mostly used in arithmetic operations. Basic applications with many examples are presented, using MATLAB® and Excel. We may call this is the fundamental part in numerical analysis.

Chapter 4, "Approach to Solve Problems by Computers," illustrates how to tackle a solution of a problem. Modeling and simulation are principal approaches for the quantitative description of chemical engineering processes in solving problems. Methodology and problem solving guidelines are described.

Basic applications with more than 15 examples are presented in Chapters 3 and 4. The solutions to these examples are obtained using MATLAB and Excel. MATLAB is a powerful code-based mathematical and engineering calculation program. It performs all calculations using matrices and vectors in a logical programming environment.

In addition, a set of application problems for Section II is included in Appendix A.

3 Numerical Methods and Chemical Engineering Computations

The assumption of an absolute determination is the essential foundation of every scientific enquiry.

(Heilbron, 2000)

One of the main objectives of Chapters 3 and 4 (Section II) is to devise, evaluate, and apply numerical techniques for employing computers to solve problems in chemical engineering. In Chapter 3, numerical methods are introduced with emphasis on elementary applications. Definitions of basic concepts are given; areas and domain of numerical methods are identified; and basic applications for many examples are presented.

3.1 INTRODUCTION

Computer scientists, programmers, and technology support staffs take a very different view of the computer from other users. They must know the details of how operating systems work and how to code various scripts that control function. They must be able to control the low-level details that a user simply assumes.

The focus of numerical methods is simply translating engineering problems into algorithms and implementing them using computers. Numerical method is a branch of applied mathematics, concerned with methods for solving complicated equations using arithmetic operations, often so complex that they require a computer, to approximate the analysis process. Numerical analysis naturally finds applications in all fields of engineering and the physical sciences. It is an important underpinning for techniques used in computational science in general. In other words, using computers to solve problems is automatically tied up to numerical techniques.

In the development of numerical methods, simplifications need to be made to progress toward a solution. For example, general functions may need to be approximated by polynomials, and computers, on the other hand, cannot generally represent numbers exactly. Therefore, numerical methods do not usually give the exact answer for a given problem at hand. However, it could be stated that the overall goal of the field of numerical analysis is the design and analysis of techniques to give approximate but accurate solutions to problems.

3.2 BASIC DEFINITIONS AND INTRODUCTORY REMARKS

The subject of numerical methods is the study of quantitative approximations to the solutions of mathematical problems including consideration of and bounds to the errors involved.

The arithmetic model used is called *algorithm*, the set of procedures the computer executes is called *program*, and the commands that carry out the procedures are called *code*.

3.2.1 Algorithms

According to Blass and Gurevich (2003), an algorithm is defined as an effective method expressed as a finite list of well-defined instructions for calculating a function. Starting from an initial state and initial input, the instructions describe a computation that, when executed, proceeds through a finite number of well-defined successive states, eventually producing *output* and terminating at a final ending state.

In computer systems, an algorithm is basically an instance of logic written in software by software developers to be effective for the intended *target* computer(s) for the target machines to produce *output* from given *input* (perhaps null).

Algorithms resemble recipes, which tell you how to accomplish a task by performing a number of steps. For example, to bake a cake the steps are as follows: preheat the oven; mix flour, sugar, and eggs thoroughly; pour into a baking pan; and so forth.

Algorithms were originally born as part of mathematics. The word *algorithm* comes from the Arabic writer *Muhammad ibn Mūsā al-Khwārizmi*, but currently the word is strongly associated with computer science.

3.2.2 Programming Language

Programming languages must provide a notational way to represent both the process and the data. Although many programming languages and many different types of computers exist, the important first step is the need to have the solution. Without an algorithm, there can be no program.

3.3 AREAS AND DOMAIN OF NUMERICAL METHODS

Numerical analysis is widely used in many chemical engineering computations. Due to the use of computers, cumbersome calculations can be done easily and a solution is reached in a shorter time. Applications include the following:

1. *Numerical solution of systems of linear equations*: This refers to solving for x in the equation $Ax = b$ with given matrix A and column vector b. Solving for $x = A^{-1}b$.
2. *Numerical solution of systems of nonlinear equations*: This refers to root-finding problems that are usually written as $f(x) = 0$, with x a vector with n

components and f(x) a vector with m components. The most important case is where n = m.
3. *Trial and error procedures*: This refers to solution of equations using the iterative convergence methods (Newton's method).
4. Problems that require numerical differentiation/integration.
5. *Optimization*: This refers to minimizing or maximizing a real-valued function f(x). The permitted values for $x = (x_1,\ldots,x_n)$ can be either constrained or unconstrained. The *linear programming problem* is a well-known and important case; f(x) is linear, and there are linear equality and/or inequality constraints on x.
6. Use computable functions p(x) to approximate the values of functions f(x) that are not easily computable or use approximations to simplify dealing with such functions. The most popular types of computable functions p(x) are polynomials, rational functions, and others.
7. Regression and correlation of a given set of data.

3.4 APPLICATIONS USING ELEMENTARY NUMERICAL METHODS

This section encompasses a set of solved examples to illustrate some of the *basic applications* of the numerical techniques. Whenever it is difficult, for example, to integrate, differentiate, or determine some specific value of a function, a computer may be called upon to numerically approximate the desired solution. In the case of integration, finding the area under a curve is a useful tool in a large number of problems in many areas of science, especially in engineering. For a curve produced by a function, you may be able to integrate the function from, say, point *a* to point *b* and calculate the area under the curve in that way. However, for curves produced from data or for curves that are produced by some complicated functions, analytical integration may not be possible. Solution could be done in a number of ways, but one easy way is to employ Excel or use MATLAB®.

Both Excel and MATLAB provide the following types of functions for performing mathematical operations and data analysis:

- Matrix manipulation and linear algebra
- Polynomials and interpolation
- Optimization
- Differentiation and integration
- Data analysis and statistics
- Ordinary differential equations
- Solution of a single nonlinear algebraic equation

By examining the following examples presented in this section, we will attempt to demonstrate the application of these principles, followed by a set of problems. This approach will introduce students to solve more advanced practical chemical engineering problems with no barrier whatsoever. Computations, using MATLAB and Excel, are demonstrated as well in Chapters 6, 7 & 8.

END-OF-CHAPTER SOLVED EXAMPLES

Example 3.1: An Algorithm, *Find Max*

Given: A list of positive numbers.
Find: The largest number on the list.
Inputs: A list L of positive numbers. This list must contain at least one number. (Asking for the largest number in a list of no numbers is not a meaningful question.)
Outputs: A number n, which will be the largest number of the list.

Algorithm

- Set max to 0.
- For each number x in the list L, compare it to max. If x is larger, set max to x.
- max is now set to the largest number in the list.
- end
- max
- else
- disp('insert a correct set of numbers!')
- end

Example 3.2: Newton–Raphson Method (Algorithm)

In calculus, Newton's method is an iterative method for finding the roots of a differentiable function (i.e., solutions to the equation). Given a function f defined over the real x and its derivative f′, we begin with a first guess x_0 for a root of the function f. Provided the function satisfies all the assumptions made in the derivation of the formula, a better approximation x_1 is $x_1 = x_0 - f(x_0)/f'(x_0)$.

Geometrically, $(x_1, 0)$ is the intersection with the x-axis of the tangent to the graph of f at $(x_0, f(x_0))$. The process is repeated as $x_{n+1} = x_n - f(x_n)/f'(x_n)$... until a sufficiently accurate value is reached.

Consider the problem of finding the square root of a number. For example, if one wishes to find the square root of 612, this is equivalent to finding the solution to $x^2 = 612$. Therefore, the function to use in Newton's method is $f(x) = x^2 - 612$ and then $f'(x) = 2x$.

With an initial guess of 10, the sequence given by Newton's method is

$$x_1 = x_0 - f(x_0)/f'(x_0) = 10 - [(10)^2 - 612]/2(10) = 35.6$$

$$\vdots$$

$$x_5 = \quad 24.73863375376 \text{ (final answer)}$$

Example 3.3: Integration

Create the function $f(x) = e^{-x^2}(\ln x)^2$.

```
fun = @(x) exp(-x.^2).*log(x).^2;
```

Evaluate the integral from x = 0 to x = Inf.

```
q = integral (fun,0, Inf)

q = 1.9475
```

Example 3.4: Differentiation

We are going to develop a MATLAB® function to calculate the *numerical derivative* of any unidimensional scalar function fun(x) at a point x_0. The function is going to have the following functionality:

Usage: D = Deriv(fun,x_0)
- fun: Name of the unidimensional scalar function
- x_0: Point of interest (scalar)
- D: Derivative of fun at x_0 (scalar)

Now, let us try our derivative function. We create a function in a separate m-file:

function y = inverse(x): Y = 1/x. And we can call it like this: *Deriv('inverse', 1)*

The result: *Expected: −1, Obtained: −1.0000*

Example 3.5: Solution Using Matrix

Solve the following system of equations for x, y, and z by using the matrix method:

$$10x + 3y + 5z = 8$$

$$8x + 2y + 2z = 7$$

$$3x + 4y - z = 6$$

Solution

$$A = [10,3,5;8,2,2;3,4,-1];$$

$$B = [8;7;6];$$

$$X = A\backslash B;$$

$$x = X(1), \quad y = x(2), \quad z = X(3)$$

Gives the result

$$x = 0.7917$$

$$y = 0.7917$$

$$z = -0.4583$$

Example 3.6

The experimental values calculated for the heat capacity of ammonia from 0 to 500 are as follows:

T (°C)	Cp (cal/g·mole·°C)
0	8.371
18	8.472
25	8.514
100	9.035
200	9.824
300	10.606
400	11.347
500	12.045

(a) *Fit the data for the following function:*

$$Cp = a + bT + CT^2 + DT^3$$

where T is in °C

(b) *Calculate the amount of heat Q required to increase the temperature of 150 mol/h of ammonia vapor from 0°C to 200°C if you know that*

$$Q = n \int_{Tin}^{Tout} Cp\, dt$$

Solution

```
T = [0,18,25,100,200,300,400,500]

Cp = [8.371,8.472,8.514,9.035,9.824,10.606,11.347,12.045]

P = Polyfit(T, Cp,3)

n = 150; syms t

Cpf = P(4) + P(3)*t + P(2)*t^2 + P(1)*t^3;

Q = n*int(Cpf,0.200)
    2.7180e+005
```

Example 3.7: Solution of a System of Linear Equations by *Excel*

$$5x + 1y + 8z = 46$$

$$4x - 2y = 12$$

$$6x + 7y + 4z = 50$$

In matrix notation, this can be written as AX = B

$$\text{with } A = \begin{bmatrix} 5 & 1 & 8 \\ 4 & -2 & 0 \\ 6 & 7 & 4 \end{bmatrix}, \quad X = \begin{bmatrix} x \\ y \\ z \end{bmatrix}, \quad B = \begin{bmatrix} 46 \\ 12 \\ 50 \end{bmatrix}$$

If A^{-1} (the inverse of A) exists, we can multiply both sides by A^{-1} to obtain $X = A^{-1}B$. To solve this system of linear equations in Excel, execute the following steps:

B6 fx {=MINVERSE(B2:D4)}

	A	B	C	D	E	F	G	H	I	J
1										
2		5	1	8			46			
3	A	4	-2	0		B	12			
4		6	7	4			50			
5										
6		-0.0303	0.197	0.0606						
7	A^{-1}	-0.0606	-0.1061	0.1212						
8		0.1515	-0.1098	-0.053						
9										
10										

G6 fx {=MMULT(MINVERSE(B2:D4),G2:G4)}

	A	B	C	D	E	F	G	H	I	J
1										
2		5	1	8			46			
3	A	4	-2	0		B	12			
4		6	7	4			50			
5										
6							4			
7						X	2			
8							3			
9										
10										

Example 3.8: Finding the Acid Concentration

Parts of a centrifugal pump are dipped in an acid–water bath to clean them. By time, the acid–water concentration decreases because of loss of solution as the parts are removed. The bath concentration should not go beyond 50%.

Starting with 90% acid–water concentration, assume that 1% of the solution is removed with every part cleaned.

Find: How many parts, n, can be cleaned in the bath?

Given: Initial concentration = 90%

Final concentration = 50%

Usage rate = 1%/part cleaned

Solution

$$n = \frac{\log(\text{initial con} + \text{min con})}{\log[1 + \text{lost\%} / 100]}$$

```
>> initial _ con = 90;

>> min _ con = 50

>> lost = 0 .01

>> n = floor(log(initial_con/min_con)/log(1 - lost)

>> n = 59
```

Example 3.9: Radioactive Decay of Polonium

Polonium has a lifetime of 140 days. Starting with 10 g today, calculate how much is left after 250 days?

Solution

Amount left = initial amount $(0.5)^{\text{time half-time}}$

```
>> initial _ amount = 10;

>> half _ life = 140; time = 250;

>> amount left = initial amount ^_0 .5 ^ (time/half life)

>> amount_left = 2.9003
```

Example 3.10

Solve the following system of equations for x, y, and z by using the matrix method

$$10x + 3y + 5z = 8$$
$$8x + 2y + 2z = 7$$
$$3x + 4y - z = 6$$

Solution

```
A = [10,3,5; 8,2,2; 3,4, - 1] ; B = [8; 7; 6];

x = X(1), y = X(2), z = X(3)

x = X(1), y = X(2), z = X(3)
```

Gives the results

```
x = 0.7917
y = 0.7917
z = -0.4583
```

Example 3.11

Estimate the average density of a water–ethanol mixture at different water compositions knowing that

Water density = 1000 kg/m^3
Ethanol density = 780 kg/m^3
Mixture density = Xwater × water density + Xethanol × ethanol density

Solution

```
Pwater = 1000;
Pethanol = 780;
Xwater = 0:.1:1
Xethanol = 1 - Xwater
Pav = Pwater*Xwater + Pethanol*Xethanol
```

```
Xwater
= 0    0.1000   0.2000   0.3000   0.4000   0.5000   0.6000
   0.7000  0.8000   0.9000  1.0000

Xethanol
= 1.0000   0.9000  0.8000  0.7000  0.6000  0.5000  0.4000
   0.3000  0.2000  0.1000  0

Answer : ρav
= 780   802   824   846   868   890   912
 934   956   978   1000
```

REFERENCES

Blass, A. and Gurevich, Y. Algorithms: A quest for absolute definitions. *Bulletin of European Association for Theoretical Computer Science* 81. http://research.microsoft.com/~gurevich/Opera/164.pdf, 2003. Retrieved July 2012.

Heilbron, J.L. *The Dilemmas of an Upright Man. Max Planck and the Fortunes of German Science, With a New Afterword*, Harvard University Press, 2000.

4 The Approach to Solve Problems by Computers

An experiment is a question which science poses to Nature and a measurement is the recording of Nature's answer.

(Planck, 1903)

The main objective of this chapter is to learn how to set up and validate mathematical models in order to solve chemical engineering problems. The implementation of models in a structured programming language such as MATLAB® or using spreadsheets is presented.

Model process development and mathematical formulation is detailed in this chapter. Applications covering numerous examples are also included.

4.1 INTRODUCTION

Learning how to solve problems is an important part of developing competency in science and engineering. It is worth noting that most engineering problems are based upon one of three underlying principles:

1. Equilibrium, force, flux, and chemical
2. *Conservation laws*: Energy and mass
3. Rate phenomena

The solution of chemical engineering problems should be an integrated part of this text, since the principles of chemical engineering are introduced along with solving numerical problems.

In attempting to solve a problem, we will demonstrate how students can attack a problem. The next proposed procedure is an attempt to be followed:

- Identify the type of problem at hand, as shown next:

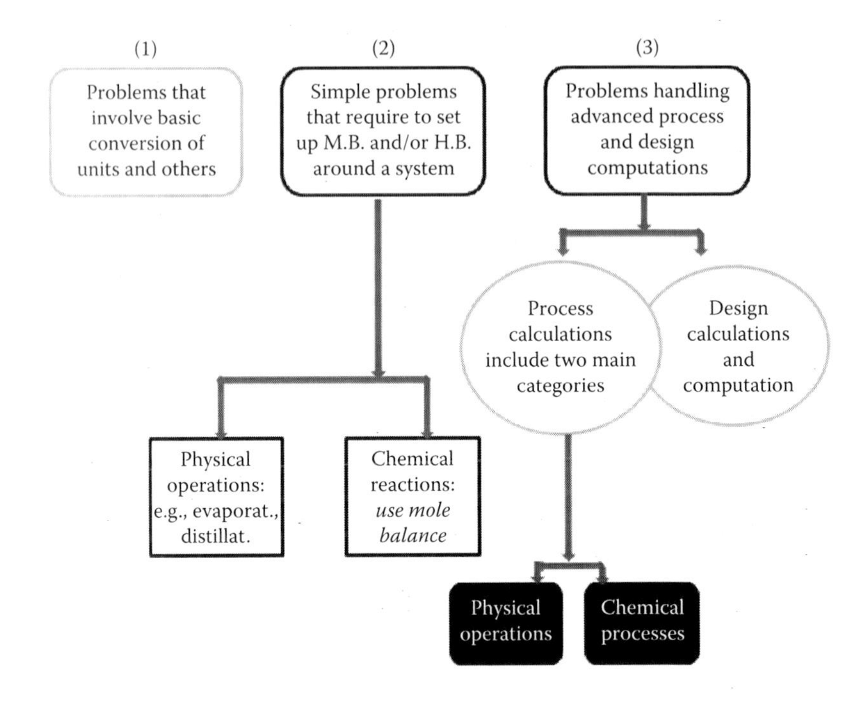

- Find out, for the problem as identified in the first procedure, if you need additional help from the sources available at your fingertip (found in the text).

4.2 METHODOLOGY

One of our main objectives in this *primer* is to demonstrate how to solve chemical engineering problems that require numerical methods by using standard algorithms, such as MATLAB or spreadsheets. To say it in simple words, it is to device and evaluate numerical techniques for employing computers to solve problems in chemical engineering.

Problem solving using computers could be handled by using

- Spreadsheets, such as Excel
- A programming language such as MATLAB

A spreadsheet such as Excel is a program that let you analyze moderately large amounts of data by placing each data point in a *cell* and then perform the same operation on groups of cells at once. *One of the advantages of spreadsheets is that data input and manipulation are relatively intuitive and hence easier than doing the same tasks in MATLAB* (Towler and Sinnott, 2013).

Details on using Excel showing how to input and manipulate data and perform operations and others are fully explained in many references found in the open literature.

Spreadsheet software has become indispensable tools in solving chemical engineering problems because of the availability of personal computers, ease of use, and adaptability to many types of problems. Handheld calculations are encouraged as well to learn a numerical technique in solving a problem.

The following basic four steps are normally involved in solving problems by computers:

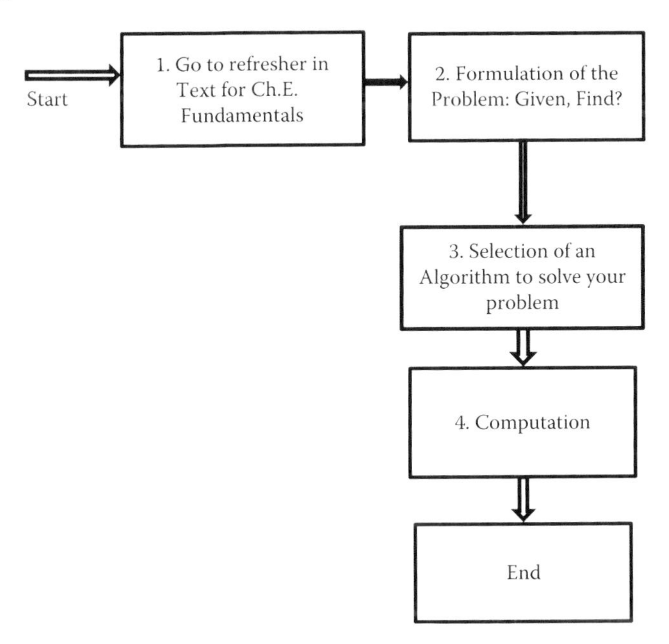

4.3 MODEL DEVELOPMENT AND MATHEMATICAL FORMULATION

The basic types of problems, we encounter in formulating mathematical modeling, fall into three main categories:

1. *Equilibrium problems*: This type of problems is recognized as steady state, where a solution does not change with time.
2. *Eigenvalue problems*: This type of problems is recognized as extensions of equilibrium problems in which critical values of certain parameters are to be determined in addition to the corresponding steady-state configuration.
3. *Transient, time-varying, or propagation problems*: This type of problems is concerned with predicting the subsequent behavior of a system from the knowledge of the initial stage.

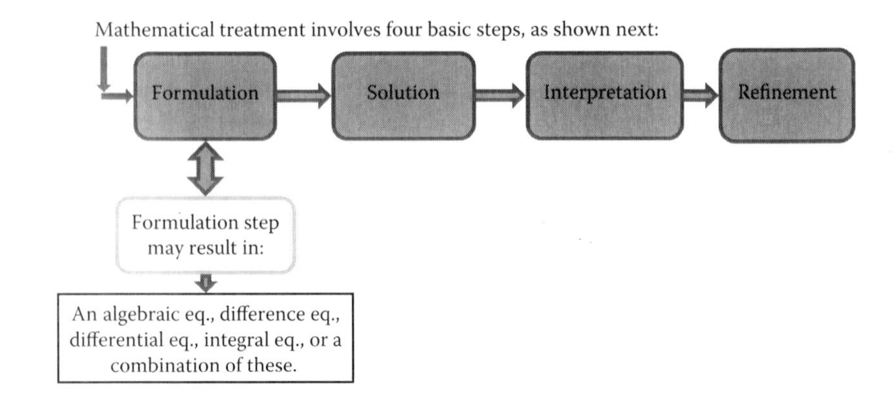

Modeling and simulation are principle approaches for quantitative description of chemical engineering processes in solving problems.

Quantitative process description is advantageous on two grounds:

1. *From a scientific point of view*: It addresses the process mechanism study, which leads to the creation of a hypothesis about the process description. This is followed by a mathematical model.
2. *From an engineering aspect*: It forms the basis of an engineering aspect of a chemical process or a chemical plant. The Association for the Advancement of Modeling and Simulation Techniques in Enterprises (AMSE) gives the following definitions:
   a. The purpose of modeling is a schematic description of the processes and the systems.
   b. The simulations are employments of the models for process investigations or process optimizations, without experiments with real systems. The introduction of interactive software packages brought about a major breakthrough in chemical engineering computations.

The solution of a problem is illustrated next:

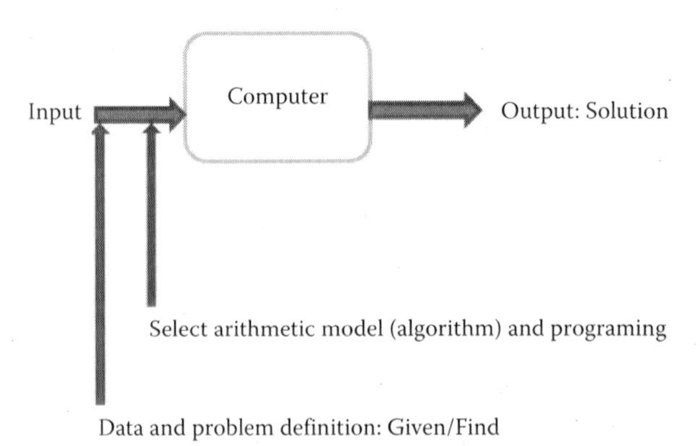

The following three steps are basically applied in this procedure:

1. Understanding the physical principles underlying the process involved in the problem in order to build a *conceptual* model.
2. Manipulation and formulation of these principles into a mathematical expression or a correlation, that is, a mathematical model. This is achieved by a thorough analysis of the engineering problems at hand, which may involve two types:
   a. Mathematical formulation (modeling) of engineering problems corresponding to specific physical situations such as momentum, heat and mass transfer, chemical reactions, and thermodynamics.
   b. Conversion of physical events and principles (e.g., a material balance) to mathematical model.
3. In engineering practice, numerical values must be incorporated and a practical solution is obtained.

These steps are presented by Figure 4.1, which illustrates different options of numerical methods involved in solving problems.

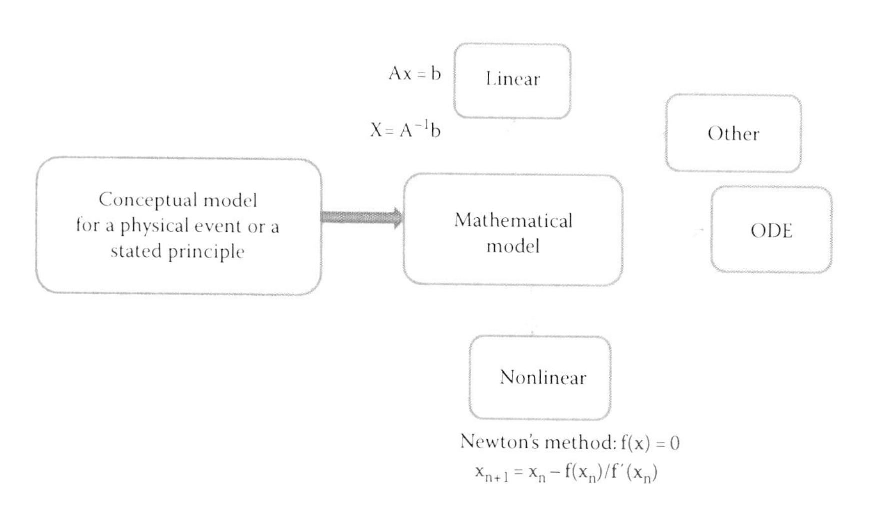

FIGURE 4.1 Formulation of mathematical model through numerical methods.

4.4 APPLICATIONS

To demonstrate the earlier procedure, the problem of flash distillation is considered. Flash calculations are very common, perhaps one of the most common chemical engineering calculations. They are a key component of simulation packages like Hysys and Aspen.

4.4.1 Problem Statement

It is required to calculate the bubble-point (BP) temperature, the dew-point temperature (DP), and the flow rates of the streams leaving the flash distillation column as well as their composition. *Mathematical formulation to determine the BP is done first through the following analysis.*

Definition: The BP temperature is physically defined as the lowest temperature at which the first bubble comes out as vapor, when the liquid is slowly heated at constant pressure. Mathematically, at the BP, the following relationships hold: (a) $\sum y_i = 1.0$ and (b) $\sum p_i = P_T$.

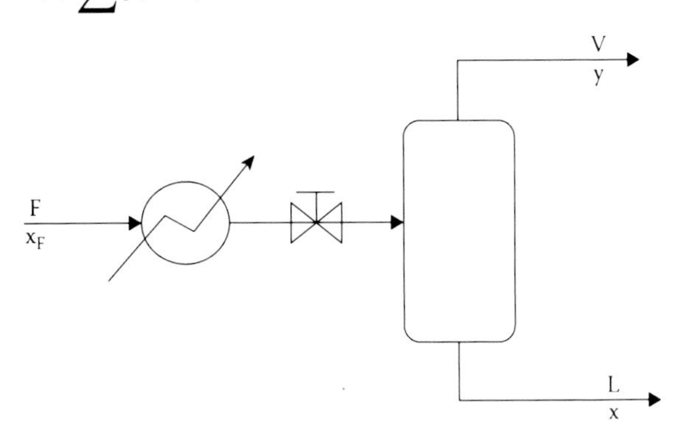

Now, mathematical formulation is pursued through these two equivalent definitions for the BP temperature. But let us first present the following fundamental relationships:

- *Raoult's law*: For a gas–liquid mixture, the partial pressure of component "i" in the liquid phase is given by

$$p_i = P_i^0 x_i \tag{4.1}$$

- *Dalton's law*: In the vapor phase, where the vapor is in equilibrium with the liquid, the partial pressure of component, i, is

$$p_i = P_T y_i \tag{4.2}$$

Take the sum of both sides of Equation 4.2:

$$\sum p_i = P_T$$

where

p_i is the partial pressure of component i
P_i^0 is the vapor pressure of pure component i
P_T is the total pressure

Equating Equations 4.1 and 4.2, we obtain

$$\frac{y_i}{x_i} = \frac{P_i^0}{P_T} = K_i \tag{4.3}$$

- Antoine's equations: The *Antoine equation* is a simple three-parameter fit to experimental vapor pressures measured over a restricted temperature range:

$$\text{Log}\, P^0 = A - \frac{B}{T + C}$$

where

A, B, and C are *Antoine coefficients* that vary from substance to substance
P is the vapor pressure of the pure component

1. *Determination of BP using the definition given in terms of the sum of y_i, by the Equation (a):* $\sum y_i = 1.0$, as shown in Figure 4.2. The following equation represents this case, where $\left[\sum y_i\right]$ is rewritten in terms of the x_i's, ($y_i = k_i x_i$):

$$f(T_{assu}) = \left[\sum k_i x_i\right] - 1 \text{ goes to zero at } T_{assu} = T_{BP} \tag{4.4}$$

2. *Determination of BP using the definition in terms of the sum of p_i, as per the Equation (b):* $\sum p_i = P_T$, as shown in Figure 4.3.

At the BP, the sum of partial pressure of the components should be equal to the total pressure on the system, P_T, or $f(T_{assu}) = \{[\sum p_i] - P_T\}$ goes to zero at $T_{assu} = T_{BP}$.

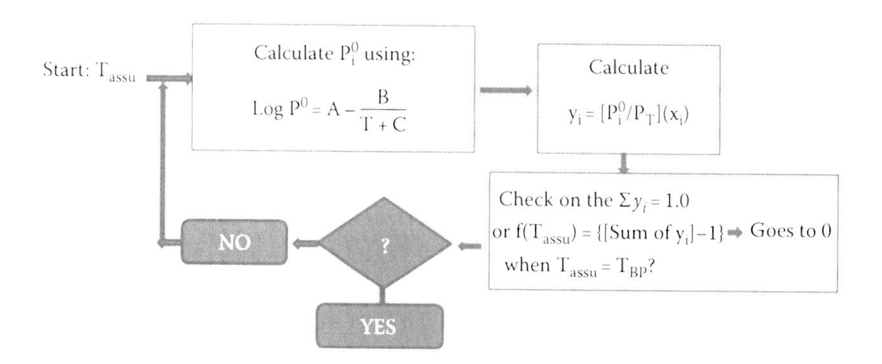

FIGURE 4.2 Determination of T_{BP} using Equation (a).

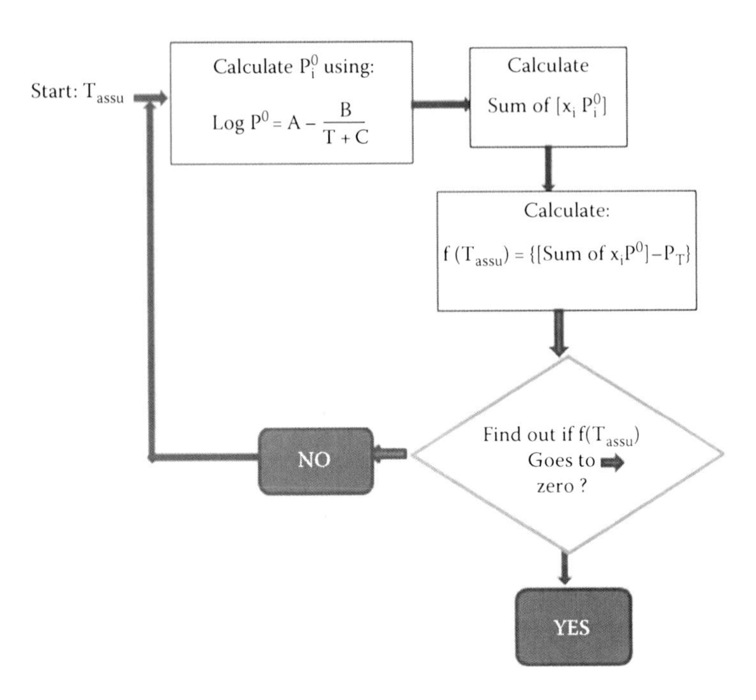

FIGURE 4.3 Determination of T_{BP} using Equation (b).

Using Equation 4.1 to replace p_i,

$$f(T_{assu}) = \left\{\left[\sum x_i P^0\right] - P_T\right\} \text{ goes to zero at } T_{assu} = T_{BP} \tag{4.5}$$

The solution of these nonlinear algebraic equations for the BP, Equations 4.4 and 4.5, using MATLAB, is presented in the following examples.

Similarly, *the DP* temperature is physically defined *as the temperature at which the first liquid drop would form when the temperature of a mixture of vapors is slowly decreased* (*cooled*) *at a specified constant* pressure.

END-OF-CHAPTER SOLVED EXAMPLES

Example 4.1

For a three-component mixture, the following information is available:

Component Number	K_i	Composition (x_i): Mole Fraction
1	$K_1 = (0.01T)/P$	1/3
2	$K_2 = (0.02)/p$	1/3
3	$K_3 = (0.03)/P$	1/3

Compute the BP temperature, T_{BP} at the specified pressure of 1 atm by using Newton's method. Take the first assumed value for T_n be equal to 100°F.

Manual Solution

Assuming T_1 = 100°F and P = 1 atm. The following calculation is carried out for the first trial:

Component	x_i	K_i at 1 atm and T = 100°F	$(K_i)(x_i)$	$[dK_i/dT]_{Tn=100}$	$(x_i)(dk_i/dT)$
1	1/3	1	1/3	0.01	(0.01)/3
2	1/3	2	2/3	0.02	(0.02)/3
3	1/3	3	3/3	0.03	(0.03)/3
Σ	1.0		6/2 = 3		0.06/3 = 0.02

From these results, it follows that

$$f(100) = [\sum K_i x_i] - 1.0 = 2 - 1 = 1$$

and

$$f'(100) = \sum x_i \frac{dK_i}{dT} = 0.02$$

where f′ stands for the first derivative of the $[\sum x_i dK_i]$ w r t T.

Applying the Newton's formula,

$$T_2 = T_1 - \left\{ \frac{f(T_1)}{f'(T_1)} \right\}$$

$$= 100 - \left(\frac{1}{0.02} \right)$$

$$= 50°F$$

Carry on one more trial to check the final answer.

Solution by Excel

B	C	D	E	F	G	H	I	J	K	L
	T_1	Component	X_i	K_i	X_iK_i	$[dK_i/dT]_{Tn}$	$(x_i)(dk_i/dT)$			
	100	1	1/3	1	1/3	0.01	1/300			
		2	1/3	2	2/3	0.02	1/150			
		3	1/3	3	1	0.03	1/100			
	Σ	6	1	6	2	0.06	0.02			

$f(100) = [\Sigma K_iX_i] - 1$	1	
$f'(100) = \Sigma X_i dK_i/dT$	0.02	
$T_2 = T_1 - [f(T_1)/f'(T_1)]$	50	^{0}F

Example 4.2

An equimolar vapor mixture of benzene and ethylbenzene is kept at 100°C.

Calculate the pressure at which the first drop of liquid will form and its composition.

Solution

This is a dew-pressure calculation. Antoine's constants are (% P in kPa and t in °C)

$$A = [13.8858, 14.0045]$$

$$B = [2788.51, 3279.47]$$

$$C = [220.79, 213.201]$$

T and the vapor mole fractions are introduced:

$$T = 100$$

$$Y = [0.5, 1-0.5]$$

$$P_{sat} = \exp\,[A - B/(T + C)]$$

$$= 180.0377 \quad 34.2488$$

The total pressure and the liquid mole fraction are calculated:

$$P = 1/\text{sum}\ (y./P_{sat})$$

$$x = P^{*}y./P_{sat}$$

$$P = 57.5498$$

$$x = 0.1598 \quad 0.8402$$

Example 4.3

[Numerical Solution of Linear equations using MATLAB®]

Case of distillation column: (*solution of n algebraic equations in n unknowns*)

A stream containing 35.0 wt.% benzene (B), 50.0% toluene (T), and the balance xylene (X) is fed to a distillation column. The overhead product from the column contains 67.3 wt.% benzene and 30.6% toluene. The bottoms product is fed to a second column. The overhead product from the second column contains 5.9 wt.% benzene and 92.6% toluene. Of the toluene fed to the process, 10.0% is recovered in the bottoms product from the second column, and 90.0% of the xylene fed to the process is recovered in the same stream.

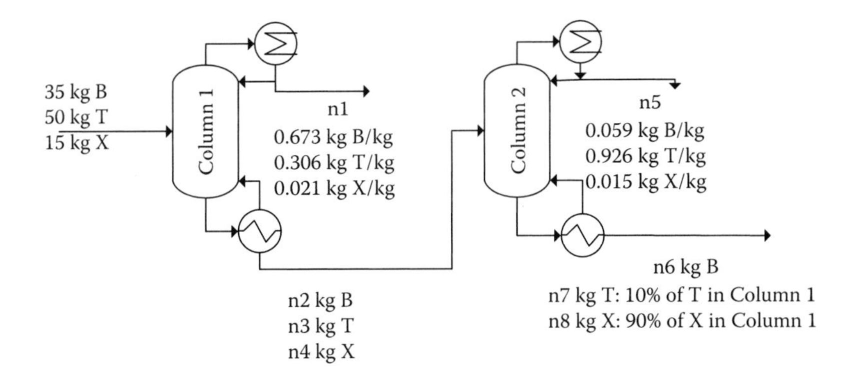

Column 1 Balances

$$B: 35.0 = 0.673n1 + n2 \tag{4.3.1}$$

$$T: 50.0 = 0.306n1 + n3 \tag{4.3.2}$$

$$X: 15.0 = 0.021n1 + n4 \tag{4.3.3}$$

Column 2 Balances

$$B: n2 = 0.059n5 + n6 \tag{4.3.4}$$

$$T: n3 = 0.926n5 + n7 \tag{4.3.5}$$

$$X: n4 = 0.015n5 + n8 \tag{4.3.6}$$

$$10\%\ T\ \text{recovery: } n7 = 0.100(50.0) = 5.00 \tag{4.3.7}$$

$$93.3\%\ X\ \text{recovery: } n8 = 0.933(15.0) = 14.0 \tag{4.3.8}$$

A solver tool can be used to solve the MATLAB equations simultaneously.

```
[n1 n2 n3 n4 n5 n6 n7 n8] = solve('35 = 0.673*n1 + n2','50 =
0.306*n2 + n3','15 = 0.021*n1 + n4','n2 = 0.059*n5 + n6','n3 =
0.926*n5 + n7','n4 = 0.015*n5 + n8','n7 = 5','n8 = 14')
```

Example 4.4

Calculate the temperature and composition of a vapor in equilibrium with a liquid that is 40.0 mole% benzene, 60.0 mole% toluene at 1 atm. Is the calculated temperature a bubble-point or dew-point temperature?

Solution

Raoult's law

$$P = x_A p_A^*(T_{bp}) + x_B p_B^*(T_{bp}) + \cdots$$

Let A = benzene and B = toluene

$$f(T_{bp}) = 0.400p^*(T_{bp}) + 0.600p^*(T_{bp}) - 760 \text{ mm Hg} = 0$$

The solution procedure is to choose a temperature, evaluate P_A^* and P_B^* for that temperature from the Antoine equation, evaluate $f(T_{bp})$ from the earlier equation, and repeat the calculations until a temperature is found for which $f(T_{bp})$ is sufficiently close to 0.

Solve using initial guess (100°C):

$$0.40 \times 10^{6.89272 - \frac{1203.5311}{T+219.888}} + 0.6 \times 10^{6.95805 - \frac{1346.773}{T+219.693}} - 760 = 0$$

```
>> T = fzero(@(T)040*10^89,272 - 12,035,311/(T+219,888) )
     +06*10^95,805 - 1,346,773/(T+219,693) ) - 760,100)

T = 95 .1460

>> pA = 040*10^89,272 - 12,035,311/(T+219,888) )

pA = 472 .5616

>> pB = 06*10^95,805 - 1,346,773/(T+219,693) )

pB = 287.4384

>> yA = pA/760

yA = 0 .6218

>> yb = pB/760

yb = 0.3782
```

The dew-point pressure, which relates to condensation brought about by increasing system pressure at constant temperature, can be determined by solving the following equation for *P*:

$$\frac{y_A P^V}{p_A^*(T_{dp})} + \frac{y_B P}{p_B^*(T_{dp})} + \cdots = 1$$

$$\frac{0.1*760}{10^{6.89272 - \frac{1203.5311}{T+219.888}}} + \frac{0.1*760}{10^{6.95805 - \frac{1346.773}{T+219.693}}} - 1 = 0$$

Solve using initial guess:

```
Tdp = fzero(@(T) (01*760)/10^89272 - 12035311/(T+219888) )
        +(01*760)/10^95805 - 1346773/(T+219693) ) - 1,50)
```

T_{dp} = 52.4354 *final answer*

Example 4.5

For the mixer shown in the following, write a code to find the values of streams A, B, and C.

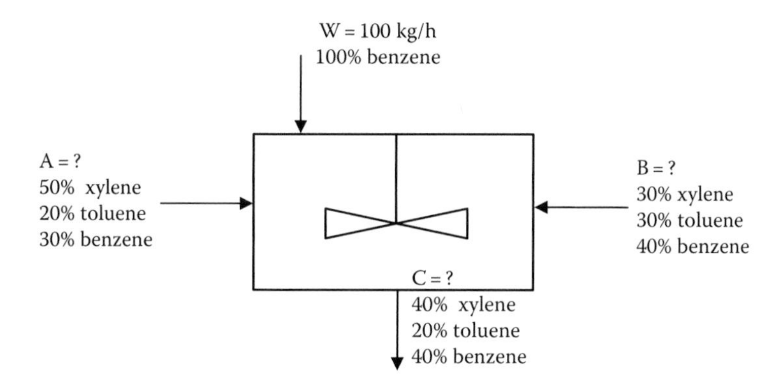

Solution

By making a component-material balance on each component within the mixer, you can reach to a system of three equations that can be solved by using the command to find the unknowns A, B, and C.

Type the following command:
`[A,B,C] = solve('.5* A + .3*B = .4*C', '.2*A + .3*B = .2*C','.3*A + .4*B + 100 = .4*C')`. The results will be the following:

```
A = 600.
B = 200.
C = 900.
```

Type the same command by entering w = 100 as the fourth equation:

```
?[A,B,C,W] = solve('.5*A + .3*B = .4*C','.2*A + .3*B =
.2*C','.3*A + .4*B + W = .4*C','W = 100')
```

The results will be the following:

```
A = 600.
B = 200.
C = 900.
W = 100.
```

Example 4.6

For the following distillation column, calculate the values of F1, F3, and F4.

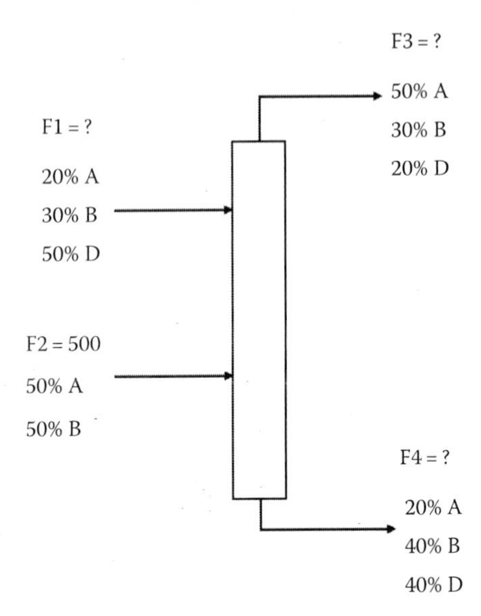


```
F1,F3,F4] = solve('.2*F1 + 250 = .5*F3 + .2*F4','.3*F1 + 250 =
.3*F3 + .4*F4','.
5*F1 = .2*F3 + .4*F4')F1 = 1000
F3 = 500
F4 = 1000
```

Example 4.7: Flash Distillation (Separation)

For the vapor liquid separater shown in the figure, write a program to calculate the values of XA, XB, YA, YB, L, and V

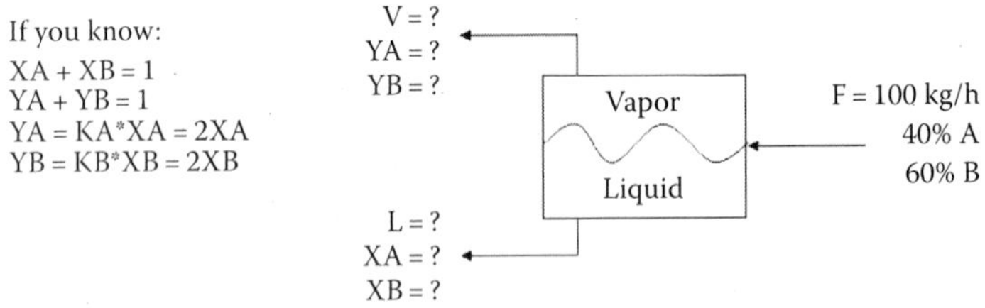

Solution:

```
A=[1,1,0,0;0,0,1,1;-2,0,1,0;0,-.5,0,1];
B=[1;1;0;0]; X=A\B;
xa=X(1),xb=X(2),ya=X(3),yb=X(4)
a=[xa,ya;xb,yb];
b=[.4*100;.6*100];
x=a\b;
L=x(1),V=x(2)
```

Gives the results

```
xa =
0.3333xb =
0.6667 ya =
0.6667 yb = 0.3333
L = 80
V = 20.0000
```

Example 4.8: Txy Diagram for Ethanol–Water System

Plot the Txy diagram for ethanol–water system, knowing the vapor pressure for three components is calculated by

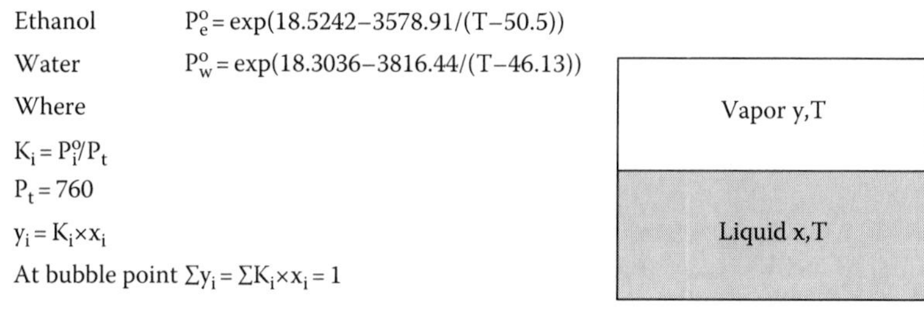

Ethanol $P_e^o = \exp(18.5242 - 3578.91/(T - 50.5))$

Water $P_w^o = \exp(18.3036 - 3816.44/(T - 46.13))$

Where

$K_i = P_i^o / P_t$

$P_t = 760$

$y_i = K_i \times x_i$

At bubble point $\Sigma y_i = \Sigma K_i \times x_i = 1$

Solution:

```
Xe = 0:.1:1;
Xw = 1-Xe;
for m = 1:11
for T = 273.15:.01:450;
```

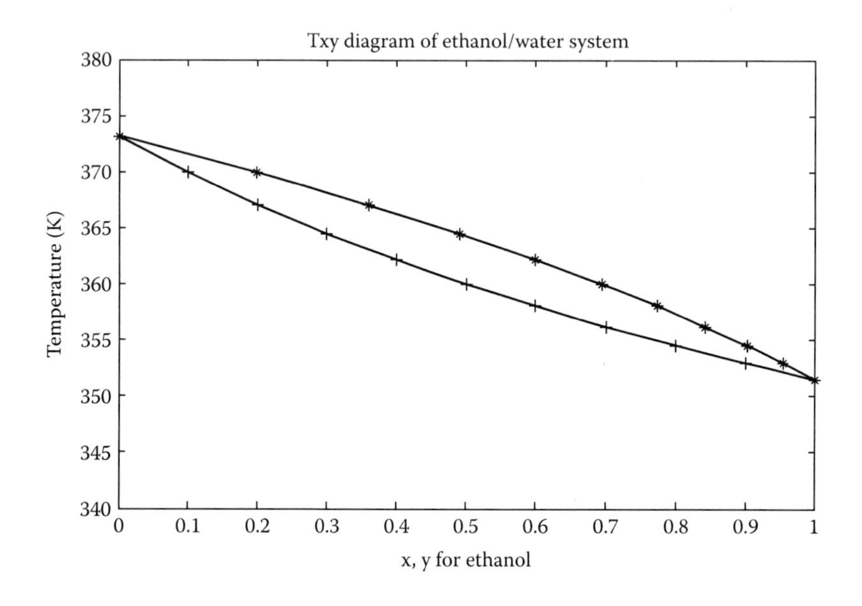

Example 4.9: Txy Diagram for Benzene–Toluene Mixture Using Excel

The main five equations relating the seven variables x, T, p_a^*, p_b^*, p_a, p_b, and y are given in the following.

Equations 4.9.1 and 4.9.2 are the Antoine equations for benzene (a) and toluene (b), respectively. These give the vapor pressure pi* in mm Hg of the pure liquids, a or b versus temperature T in °C. Here, x_a is the mole fraction of benzene in the liquid, y_a is its mole fraction in the vapor, and p_i is the partial pressure of i for a or b. Equations 4.9.3 and 4.9.4 come from assuming an ideal liquid mixture, which is approximately true for benzene and toluene.

Find

The goal is to create a Txy phase diagram for mixtures of benzene and toluene, where T is the temperature, x is the mole fraction of benzene in the liquid, and y is the mole fraction of benzene in the vapor. A horizontal line drawn for a given T gives the compositions of liquid and vapor in equilibrium at that T. Such diagrams are very useful for distillation calculations

Approach: Using these equations, you will create a Txy graph for benzene–toluene mixtures at 1 atm. This graph will show the mole fraction ya of benzene in the vapor corresponding to equilibrium with liquid of mole fraction xa at temperature T required to give a total pressure of 760 mm Hg:

$$\log_{10} p_a^* = 6.814 - \frac{1090}{197.1 + T} \tag{4.9.1}$$

$$\log_{10} p_b^* = 7.136 - \frac{1457}{231.8 + T} \tag{4.9.2}$$

$$p_a = x p_a^* \tag{4.9.3}$$

$$p_b = (1 - x) p_b^* \tag{4.9.4}$$

$$y = \frac{p_a}{p_a + p_b} \tag{4.9.5}$$

Proposed Procedure

1. Create an Excel spreadsheet giving T and y versus x for P = 760 Torr (mm Hg), and then execute the following steps:
    a. Enter the values for the Antoine constants from Equations 4.9.1 and 4.9.2.
    b. In column A enter values for x from 0 to 1 by increments of 0.1.
    c. Leave column B blank for values of T.
    d. In column C calculate p_a^* using the Antoine equation, the constants for benzene, and the temperature in column B.

e. In column D calculate p_b^* using the Antoine equation, the constants for toluene, and the temperature in column B.
f. In column E calculate $P = p_a + p_b$ using columns C and D and Equations 4.9.3 and 4.9.4.
g. In column F use Equation 4.9.5 to calculate the values of y.
h. Use Goal Seek row-by-row to calculate T required to give P = 760 mm Hg. This will automatically fill in the correct values in all columns.

2. Use the results in the spreadsheet to create a graph, with x and y on the horizontal (X) axis and T on the vertical (Y) axis.
3. Format the graph.

Results will be as shown in Figure 4.4:

	A	B	C	D	E	F
3	For total pressure P = 760 Torr (1 atm)					
4						
5	Antoine equation coefficients					
6	Compound	A	B	C		
7	Benzene	6.814	1090	197.1		
8	Toluene	7.136	1457	231.8		
9					P	
10	x	T (°C)	p_a^* (Torr)	p_b^* (Torr)	$(p_a + p_b)$	y
11	0.0	110.6	1869.2	760.0	760.0	0.000
12	0.1	105.8	1644.0	661.8	760.0	0.216
13	0.2	101.6	1463.6	584.1	760.0	0.385
14	0.3	97.9	1316.6	521.5	760.0	0.520
15	0.4	94.6	1194.9	470.1	760.0	0.629
16	0.5	91.6	1092.7	427.3	760.0	0.719
17	0.6	88.9	1005.8	391.3	760.0	0.794
18	0.7	86.4	931.2	360.5	760.0	0.858
19	0.8	84.1	866.5	334.0	760.0	0.912
20	0.9	82.0	809.9	311.0	760.0	0.959
21	1.0	80.0	760.0	290.8	760.0	1.000

FIGURE 4.4 Calculation of the equilibrium data for Benzene/Toluene mixture using Excel. (Courtesy of Clarkson University, Engineering spreadsheets Using Excel II. Goalseek and Graphing, Potsdam, NY, n.d., http://clarkson.edu/~wwilcox/ES100/xl-tut2.pdf, accessed October 14, 2014.)

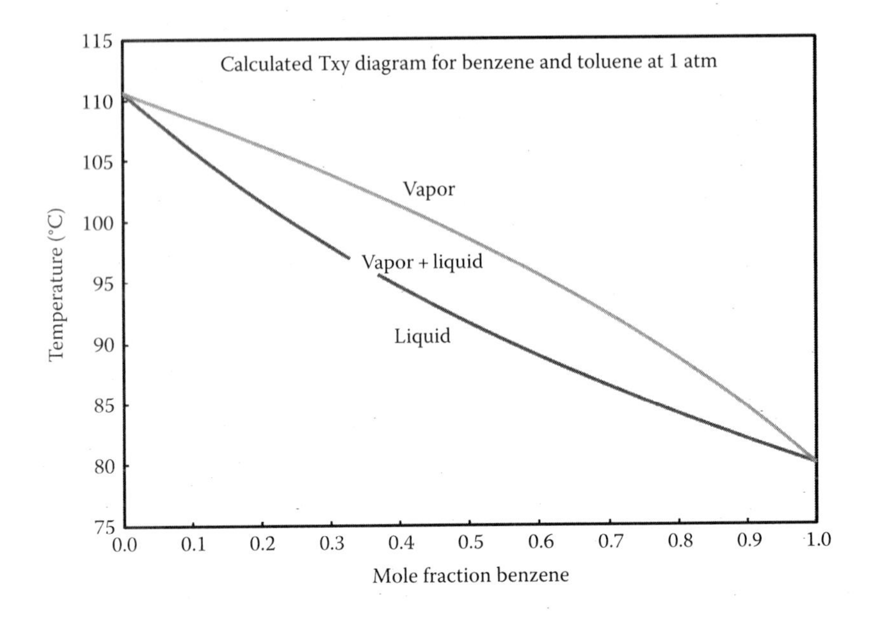

Example 4.10

Given

i. The vapor pressure data measured at a pressure of 101.32 kPa as shown in Table 4.1
ii. Using Raoult's law

Find: Calculate the vapor and liquid compositions in equilibrium (in mole fractions, y and x) for the benzene–toluene system, and plot the Txy diagram.

TABLE 4.1
Temperature–Vapor Pressure Data for Benzene–Toluene at 101.32 kPa

Temperature		Vapor Pressure			
		Benzene		Toluene	
K	°C	kPa	mm Hg	kPa	mm Hg
353.3	80.1	101.32	760	—	—
358.2	85.0	116.9	877	46.0	345
363.2	90.0	135.5	1016	54.0	405
368.2	95.0	155.7	1168	63.3	475
373.2	100.0	179.2	1344	74.3	557
378.2	105.0	204.2	1532	86.0	645
383.3	110.6	240.0	1800	101.32	760

Solution

Step 1

Write Raoult's law for a binary mixture of A and B:

$$P_A = P_A^* x_A; \quad P_B = P_B^* x_B = P_B^*(1 - x_A)$$

The total pressure PT = PA + PB

Replacing for the partial pressures and rearranging, we have

$$P_T = P_A^* x_A + P_B^*(1 - x_A)$$

$$P_T = (P_A^* - P_B^*)x_A + P_B^*$$

Rearranging, we have the expression for x_A

$$x_A = \frac{P_T - P_B^*}{P_A^* - P_B^*}$$

Since $P_A^* x_A = y_A P_T$, we have the expression for y_A

$$y_A = \left(\frac{P_A^*}{P_T}\right) x_A$$

Step 2

The vapor-liquid-equilibrium (VLE) data can be calculated at each temperature by substituting for total pressure (101.32 kPa) and the appropriate vapor pressures. For example, at 85.0°C,

$$x_A = \frac{101.32 - 46.0}{116.9 - 46.0} = 0.7803$$

$$y_A = \left(\frac{116.9}{101.32}\right)(0.7803)$$

Step 3

Repeating for other temperatures using *Excel*, the following results are obtained

Temperature (°C)	Mole Fraction of Benzene in Liquid (x)	Mole Fraction of Benzene in Vapor (y)
80.1	1.000	1.000
85.0	0.780	0.900
90.0	0.581	0.777
95.0	0.411	0.632
100.0	0.258	0.456
105.0	0.130	0.261
110.6	0.000	0.000

Source: Chemical Eng. Computing Lab. Manual, KFUPM, Dhahran, Saudi Arabia, September 2001.

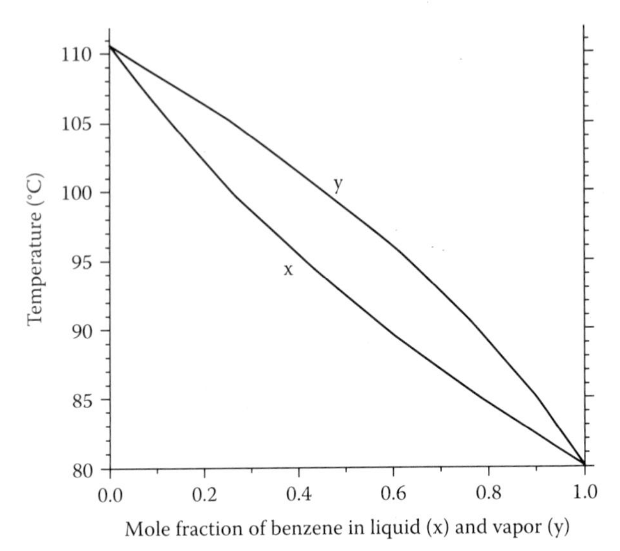

Txy diagram for benzene-toluene system at 1 atm

Example 4.11

Plot Pxy diagram for the binary system of acetonitrile (1)/nitromethane (2). Vapor pressures for the pure species are given by the following equations:

For acetonitrile (1), $P_1^o = \exp(14.2724 - 2945.47/(T + 224))$.

For nitromethane (2), $P_2^o = \exp(14.2043 - 2972.64/(T + 209))$,
in which
$T = 75°C$
P_1 and P_2 in kPa

$$K_i = \frac{P_i^o}{P_t}$$

At bubble point, $\sum y_i = \sum K_i \times x_i = 1$

Solution

Write the following code:

```
X1 = 0:.1:1;
X2 = 1-X1;
T = 75;
P1 = exp(14.2724 - 2945.47/(T + 224));% Vapor pressure of
acetonitrile
P2 = exp(14.2043 - 2972.64/(T + 209));% Vapor pressure of
nitromethane for m = 1:11
for Pt = .1:.01:120;
K1 = P1/Pt; % acetonitrile
K2 = P2/Pt; % nitromethane
```

```
sum = K1*X1(m) + K2*X2(m);
if sum<1
break
end
end
Press(m) = Pt;
Y1(m) = K1*X1(m);
end
plot(X1,Press,'k-+',Y1,Press,'k-*')
axis ([0 1 20 100])
xlabel('x,y for acetonitrile')
ylabel('Pt (kpa)')
title('Pxy diagram of system acetonitrile/nitromethane
system')
```

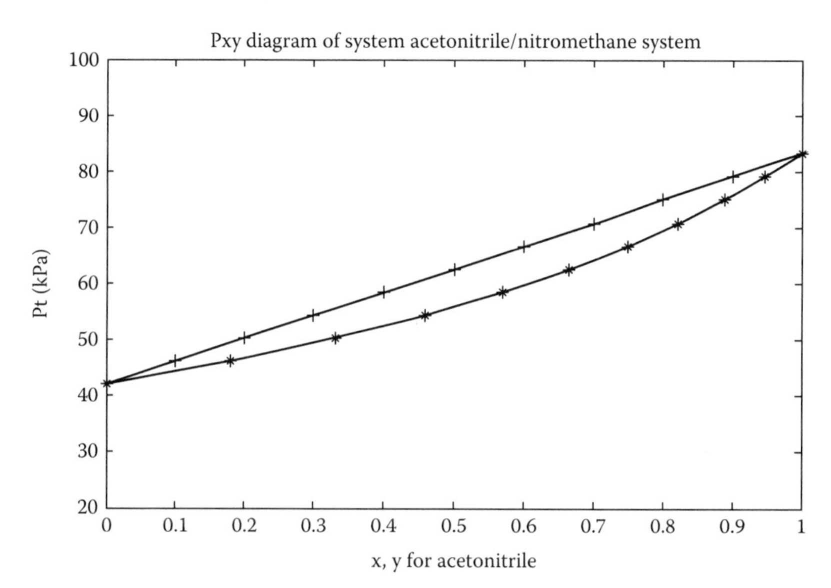

REFERENCES

Clarkson University, n.d., Engineering Spreadsheets Using Excel II. Goalseek and Graphing, Potsdam, NY. http://clarkson.edu/~wwilcox/ES100/xl-tut2.pdf (accessed October 14, 2014).

Planck, M. *Treatise on Thermodynamics*, A. Ogg (Trans.), London: Longmans, Green & Co. https://archive.org/details/treatiseonthermo00planrich, 1903.

Towler, G. and Sinnott, R.K. *Chemical Engineering Design: Principles, Practice and Economics of Plant and Process Design*, New York: Elsevier, 2013.

Section III

Backbone Materials

As stated in the "Introduction", a handy collective source of basic principles and techniques in chemical engineering organized in a summary manner is a prime feature of this text. Having presented—in Sections I and II—the basic concepts and principles as introduction to calculations and described how to device and evaluate numerical techniques to solve problems through standard algorithm, we turn next to Section III.

Section III consists of Chapters 5 through 10. They represent what may be named "The Backbone Materials of Chemical Engineering." Each chapter in this section, except Chapter 10, will encompass two sections: one includes the theoretical principles in a condensed format, the other presents a set of numerically solved problems using interactive numerical software packages (Excel, MATLAB®). This way, the dual themes of presenting the fundamentals followed by computer applications are realized. Application problems for Section III are found in Appendix A.

Chapter 10 stands unique in its function. It includes six case studies covering different topics.

5 Fluid Flow and Transport of Fluids

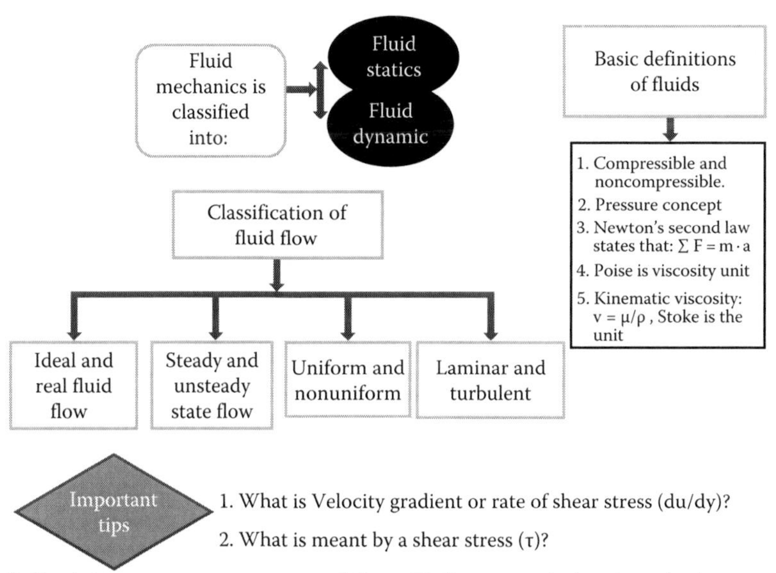

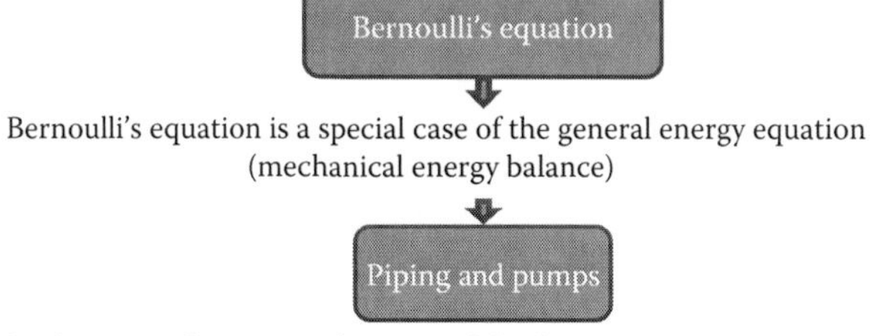

3. Absolute pressure = gauge pressure (taken with the proper sign) + atmospheric pressure.
4. Fluid statics is concerned with the static properties and behavior of fluids

Bernoulli's equation

Bernoulli's equation is a special case of the general energy equation (mechanical energy balance)

Piping and pumps

1. The diameter of a pipe is a function of the flow rate of the fluid: D = f (Q), while
2. The thickness of a pipe (t) is a function of the working pressure inside the pipe: t = f (p)

5.1 INTRODUCTION

Before a formal definition of fluid is given, it is known that the three common states of matter are solid, liquid, and gas. A fluid is either a liquid or a gas. A *fluid* is defined as a substance that deforms continuously under the application of a shear stress.

The definition of a stress would illustrate some basic principles. A *stress* is defined as a force per unit area, acting on an infinitesimal surface element. Stresses have both *magnitude* (force per unit area) and *direction*, and the direction is relative to the surface on which the stress acts. Stresses are classified into two main types: *normal* stresses and *tangential* stresses. *Pressure* is an example of a *normal* stress and acts inward, toward the surface, and perpendicular to the surface; on the other hand, a *shear* stress is an example of a *tangential* stress, that is, it acts along the surface, parallel to the surface. Friction due to fluid viscosity is the primary source of shear stresses in a fluid.

Fluid mechanics has two branches, as shown next:

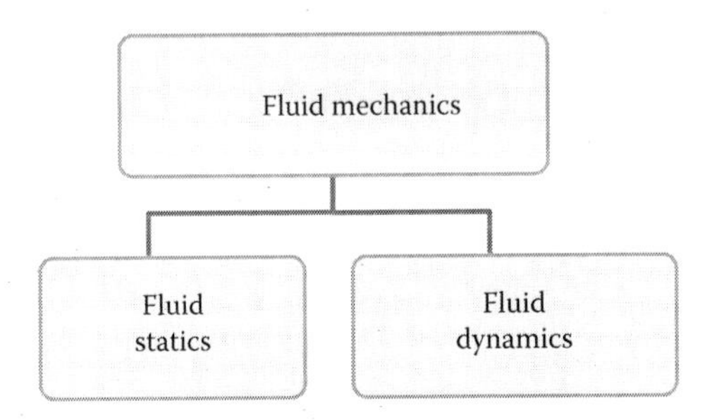

Fluid statics treats fluid in the equilibrium state (no motion), while fluid dynamics treats fluids when portion of the fluid are in motion (concerned with the relation between the fluid velocity and the forces acting on it).

5.2 DEFINITIONS AND TERMINOLOGY IN FLUID MECHANICS

5.2.1 Compressible and Noncompressible Fluids

A compressible fluid is identified by a fluid density sensitive to any change in temperature or pressure (including gases and vapors). On the other hand, if no change or little change in density occurs with change of pressure or temperature, the fluid is termed incompressible fluid (including liquids).

5.2.2 Pressure Concept

The basic property of a static fluid is pressure. Pressure is familiar as a surface force exerted by a fluid against the walls of its container.

5.2.3 Forces Acting on a Fluid

Any fluid may be subjected to three types of forces:

1. Gravity force (body force: acts without physical contact)
2. Pressure force (surface force: requires physical contact for transmission)
3. Shear force (appears in case of dynamic fluids) (surface force)

Newton's second law states that

$$\sum F = m \cdot a$$

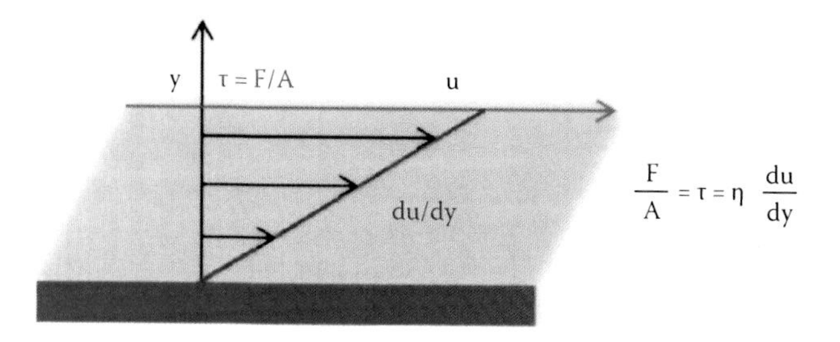

FIGURE 5.1 Representation of viscosity.

This law is applied for fluid statics and fluid dynamics, while for fluid statics,

$$\sum F = 0.0$$

5.2.4 Nature of Fluids

As explained earlier, a fluid undergoes continuous deformation when subjected to a shear stress. The resistance offered by a *real fluid* to such deformation is called its *consistency*. Now, if the static pressure and temperature are fixed, the consistency of the fluid is constant. This is defined as a *Newtonian* fluid (simple fluids and gases), and consistency is called *viscosity*. If, however, consistency is a function of shear stress as well as of temperature, the fluid is called *nonnewtonian* fluid (see Figure 5.1).

The unit of viscosity (i.e., absolute viscosity) in the cgs system is defined as *the poise* (p) = 1 g/(cm)(s). Viscosities are usually tabulated in *centipoises*: 1 (cp) = 0.01 poise.

Kinematic viscosity of a fluid of density ℓ lb/cu · ft and viscosity μ lb/(ft)(s) is given by

$$v = \frac{\mu}{\rho} \text{ ft}^2/\text{s}$$

The cgs unit of kinematic viscosity is known as *stoke*:

$$1 \text{ stoke} = 1 \text{ cm}^2/\text{s}$$

5.3 CLASSIFICATION OF FLUID FLOW

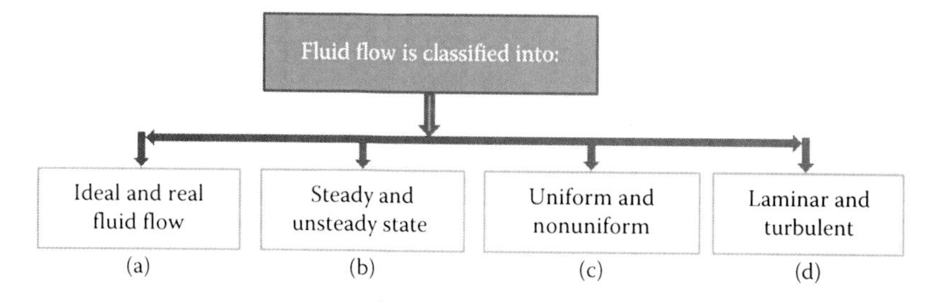

5.3.1 Ideal Fluid Flow

This flow is characterized by the following:

- There is no friction (viscosity is zero), that is, there is no dissipation of mechanical energy into heat.
- All particles flow in parallel lines and equal velocities (no velocity gradient).
- There is no formation of eddies or circulation within the stream.

This type of flow is also called potential flow or irrotational flow. This can exist at a distance not far from a solid boundary (outside the boundary layer).

5.3.2 Real Fluid Flow

This type of flow is characterized by the following:

- There is the presence of friction.
- There is a velocity gradient.

This type exists inside the boundary layer where the fluid is affected by the presence of solid boundaries.

5.3.3 Steady and Unsteady State Flow

- *Steady state*: In this type of flow, the conditions are independent of time (invariant with time).
- *Unsteady state*: The conditions are dependent to time (change with time).

5.3.4 Uniform and Nonuniform Flow

In the uniform type of flow, the conditions (velocity) are independent on position (space coordinate), while the conditions for nonuniform flow are position dependent. It is to be noted that uniform flow is ideal flow and nonuniform flow is real flow.

5.3.5 Laminar Flow

This type of flow exists at low velocities and assumes that the fluid adjacent layers slide past one another like playing cards. This type of flow is characterized by the fact that there is no lateral mixing and no crosscurrent or eddies, and the velocity gradient is high.

5.3.6 Turbulent Flow

Turbulent flow exists at high velocities and is characterized by the fact that there is mixing and crosscurrents, and the velocity gradient is lower than that of laminar flow.

5.4 PARAMETERS IN LAMINAR FLOW

Two important parameters are identified in laminar flow as explained next.

5.4.1 Velocity Gradient or Rate of Shear Stress (du/dy)

For a steady-state 1D flow of an incompressible fluid over a solid plane surface, the picture is as presented in Figures 5.2 and 5.3. Now, by plotting the velocity versus distance (in y direction), we observe the following:

- The velocity is zero at the wall.

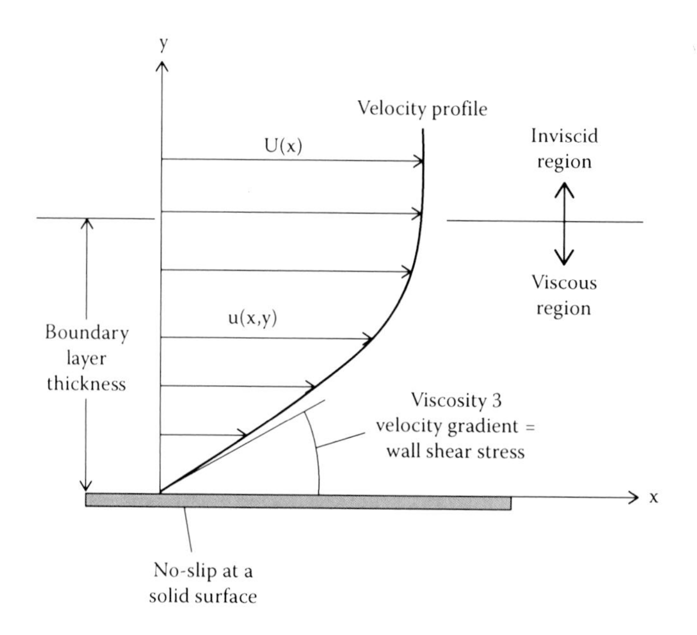

FIGURE 5.2 Velocity gradient.

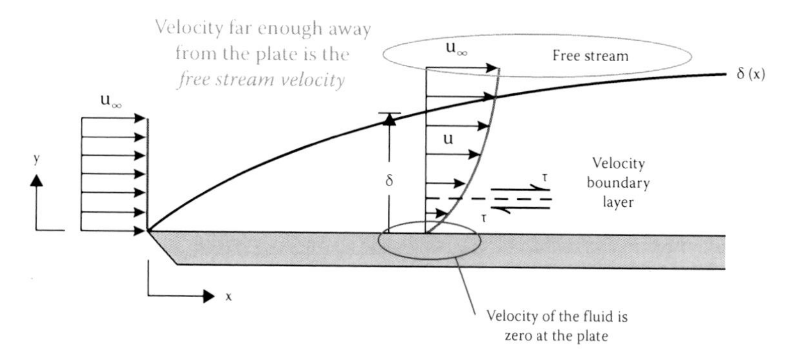

FIGURE 5.3 Change of velocity profile with x.

- As the distance increases, the velocity increases (with a decreasing rate) until the maximum velocity is reached after which the fluid will be not affected by the wall.
- The fluid velocity at which the fluid is not affected by the wall is called the *free stream velocity.*

By plotting (y) versus (du/dy), the figure shows that the velocity gradient (rate of shear stress) is maximum at the wall and decreases as the distance increases to reach the minimum value at the free stream velocity.

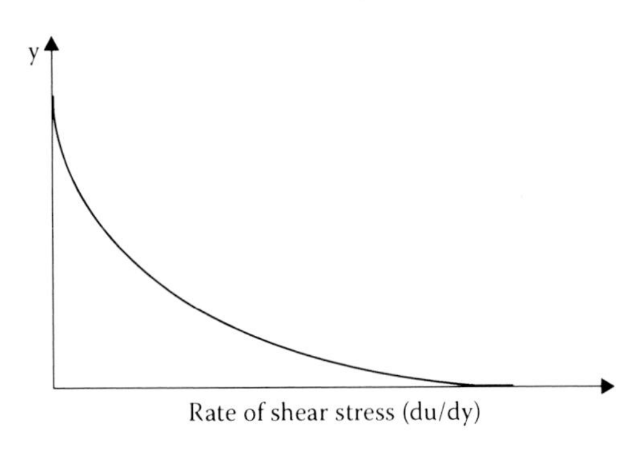

5.4.2 Shear Stress (τ)

Wherever there is a velocity gradient, a shear force must exist. The shear force acts parallel to the plane of the shear. The ratio between the shear force to the shear area is called the shear stress.

5.5 FLUID STATICS

Fluid statics is concerned with the static properties and behavior of fluids. Two cases are known:

1. Hydrostatics, the case dealing with liquids
2. Pneumatics, the case dealing with gases

A body of liquid in a static equilibrium is being acted upon only by compressive forces. The intensity of this force is known as the static pressure. It is expressed in terms of

- Pound force per square inch
- Dyne per square meter

5.5.1 Gauge Pressure and Absolute Pressure

The difference between a given fluid pressure and the atmospheric pressure is known as the *gauge pressure*. Reading of pressure gauges is normally positive gauge pressures. *Absolute pressure*, on the other hand, is the true total pressure and is given by

Absolute pressure = Gauge pressure (taken with the proper sign)
+ atmospheric pressure

The term *static* head generally means the pressure of a fluid exerted by the head of a fluid above the point in question.

Newton's law, which is given by the equation force = mass × acceleration, is used to determine the pressure as shown.

For constant density fluids, the static head is given by

$$p_h = \frac{h\rho g}{g_c}$$

where

h is the head of liquid above the point, ft
ρ is the liquid density, lb/cu · ft
g is the local acceleration due to gravity, ft/s²
g_c is the dimensional constant, 32.17 (lb)(ft)/(lb · force)(s²)

5.6 OVERALL ENERGY BALANCE EQUATION

5.6.1 Introduction

In order to write an energy balance, we need to know what kinds of energy can enter or leave a system. Energy is often categorized as

- Kinetic energy (KE)
- Potential energy (PE)
- Internal energy (U)

The total amount of energy entering the system is the sum of all of the different types entering the system. Energy may be transferred between a system and a boundary by two modes:

1. Heat: Q
2. Work: W

5.6.2 General Energy Balance Equation

Now, assume a system having a total energy E. If Q is added to it from the surrounding, then as a result, W, is done by the system, and the output streams are leaving. Then

Energy available in output streams = Energy available in input streams
that is,

$$\Delta U + \Delta K.E. + \Delta P.E. = Q - W_T \text{ (first law of thermodynamics)}$$

or

$$\Delta U + \Delta K.E. + \Delta P.E. + \Delta PV = Q - W_s \tag{5.1}$$

where W_T is the sum of $W_s + \Delta PV$. This term, ΔPV, is called the *flow work* that is the work done on the fluid at the inlet minus work done by the fluid at the outlet.

If we define

$$\Delta U + \Delta PV = \Delta H,$$

enthalpy changes and the following equation is obtained:

$$\Delta H + \Delta K.E. + \Delta P.E. = Q - W_s = \Delta E = \text{Change in total energy of a system} \tag{5.2}$$

5.6.3 Special Cases

5.6.4 Mechanical Energy Balance

Replacing V by $1/\rho$ in Equation 5.1, we get

$$\Delta K.E. + \Delta P.E. + \frac{\Delta P}{\rho} + (\Delta U - Q) = -W_s \tag{5.3}$$

This term, $\Delta U - Q$, represents the friction losses in our system. It means that this energy appears as friction losses, F. Equation 5.3 can be rewritten as

$$\Delta K.E. + \Delta P.E. + \frac{\Delta P}{\rho} + F + W_s = 0 \tag{5.4}$$

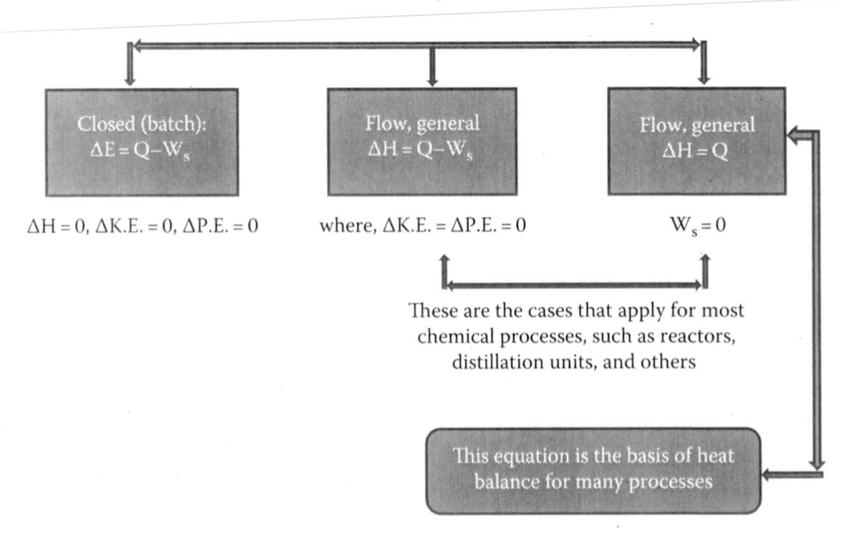

5.6.5 Bernoulli's Equation

Bernoulli's equation is a special case of the general energy equation (mechanical energy balance) that is probably the most widely used tool for solving fluid flow problems. It provides an easy way to relate the elevation head, velocity head, and pressure head of a fluid.

Bernoulli's equation results from the application of the general energy equation and the first law of thermodynamics to a steady-flow system.

The general form of the energy balance equation is

$$\Sigma(\text{all energies in}) = \Sigma(\text{all energies out}) + \Sigma(\text{energy stored in system})$$

$$\Sigma \text{Ein} = \Sigma \text{Eout} + \Sigma \text{Estorage}$$

Consider a steady-flow system in which *no work* is done on or by the fluid, *no heat* is transferred to or from the fluid, and *no change occurs in the internal energy* (i.e., no temperature change) of the fluid. Under these conditions, the general energy equation is simplified to

$$(\text{PE} + \text{KE} + \text{PV})_1 = (\text{PE} + \text{KE} + \text{PV})_2$$

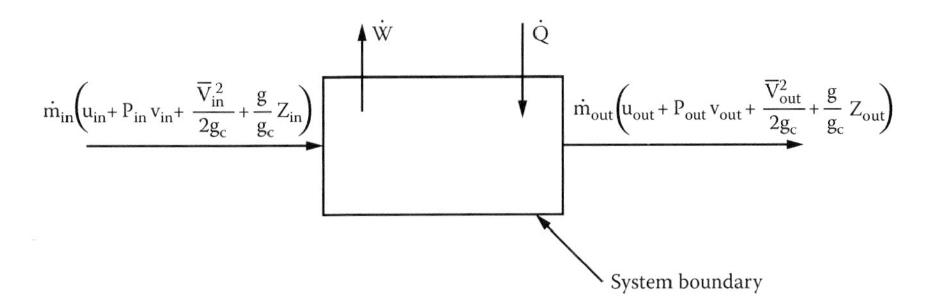

Substitution for each term

$$\frac{p_1}{\rho} + Z_1\frac{g}{g_c} + \frac{u_1^2}{2g_c} = \frac{p_2}{\rho} + Z_2\frac{g}{g_c} + \frac{u_2^2}{2g_c}$$

This equation is the simplified form of Bernoulli's equation and is used for frictionless flow; no work is applied on the fluid and no heat is added or lost from the fluid.

Each term in Bernoulli's equation represents a form of energy possessed by a moving fluid (potential-, kinetic-, and pressure-related energies). In essence, the equation physically represents a balance of the KE, PE, and PV energies so that if one form of energy increases, one or more of the others will decrease to compensate and vice versa. It is to be noted that each term of Bernoulli's equation has a unit of energy per unit mass (J/kg or $\text{lb}_f \cdot \text{ft/lb}_m$).

5.6.6 Head Form of Bernoulli's Equation

Multiplying all terms in Bernoulli's equation by the factor g_c/g results in the form of Bernoulli's equation shown

$$\frac{p_1}{\rho}\frac{g_c}{g}+Z_1+\frac{u_1^2}{2g}=\frac{p_2}{\rho}\frac{g_c}{g}+Z_2+\frac{u_2^2}{2g}$$

The units for all the different forms of energy in this equation are measured in units of distance; these terms are sometimes referred to as *heads* (pressure head, velocity head, and elevation head). Each of the energies possessed by a fluid can be expressed in terms of head.

It has been established that the elevation head represents the PE of a fluid due to its elevation above a reference level. The velocity head represents the KE of the fluid. It is the height in feet that a flowing fluid would rise in a column if all of its KE were converted to PE. The pressure head represents the flow energy of a column of fluid whose weight is equivalent to the pressure of the fluid.

The sum of the elevation head, velocity head, and pressure head of a fluid is called the *total head*.

Thus, Bernoulli's equation states that the total head of the fluid is constant.

Bernoulli's equation can be modified to take into account the friction losses in the fluid flow to be in the form:

$$\frac{p_1}{\rho}+Z_1\frac{g}{g_c}+\frac{V_1^2}{2\alpha_1 g_c}=\frac{p_2}{\rho}+Z_2\frac{g}{g_c}+\frac{V_2^2}{2\alpha_2 g_c}+\sum F$$

where F is the friction loss in the piping system between point 1 and point 2 including both the skin and form friction.

If the flow line contains a device that adds work to the fluid, Bernoulli's equation will take the form

$$\frac{p_1}{\rho}+Z_1\frac{g}{g_c}+\frac{V_1^2}{2\alpha_1 g_c}+\eta W=\frac{p_2}{\rho}+Z_2\frac{g}{g_c}+\frac{V_2^2}{2\alpha_2 g_c}+\sum F$$

where

W is the mechanical work done by the pump per unit mass of fluid

η is the pump efficiency

$$\eta=\frac{W-\text{friction losses in the pump}}{W}$$

5.6.7 Pressure Drop and Friction Losses

The Fanning friction factor is a dimensionless number used in fluid flow calculations. It is a common parameter used in laminar and especially in turbulent flow. It is defined as the drag force per wetted unit surface area (i.e., shear stress), divided by the product of density times velocity head or $\frac{1}{2}\rho\upsilon^2$. The force is Δp_f, times the cross-sectional area πR^2 and the wetted surface area $2\pi R\Delta L$.

The relation between the Δp_f due to friction and f is given as follows:

$$\Delta p_f = 4f\rho\left[\left(\frac{\Delta L}{D}\right)\left(\frac{\upsilon^2}{2}\right)\right]$$

or

$$F_f = \frac{\Delta p_f}{\rho} = 4f\left(\frac{\Delta L}{D}\right)\left(\frac{\upsilon^2}{2}\right)$$

For laminar flow, the friction factor is given by

$$f = \frac{16}{N_{Re}}$$

This relationship holds up to a Reynolds (Re) number of 2100. Beyond that and for turbulent flow and for design purposes, Moody chart, shown in Figure 5.4, is used to predict the value of f and, hence, the frictional pressure drop of round pipes.

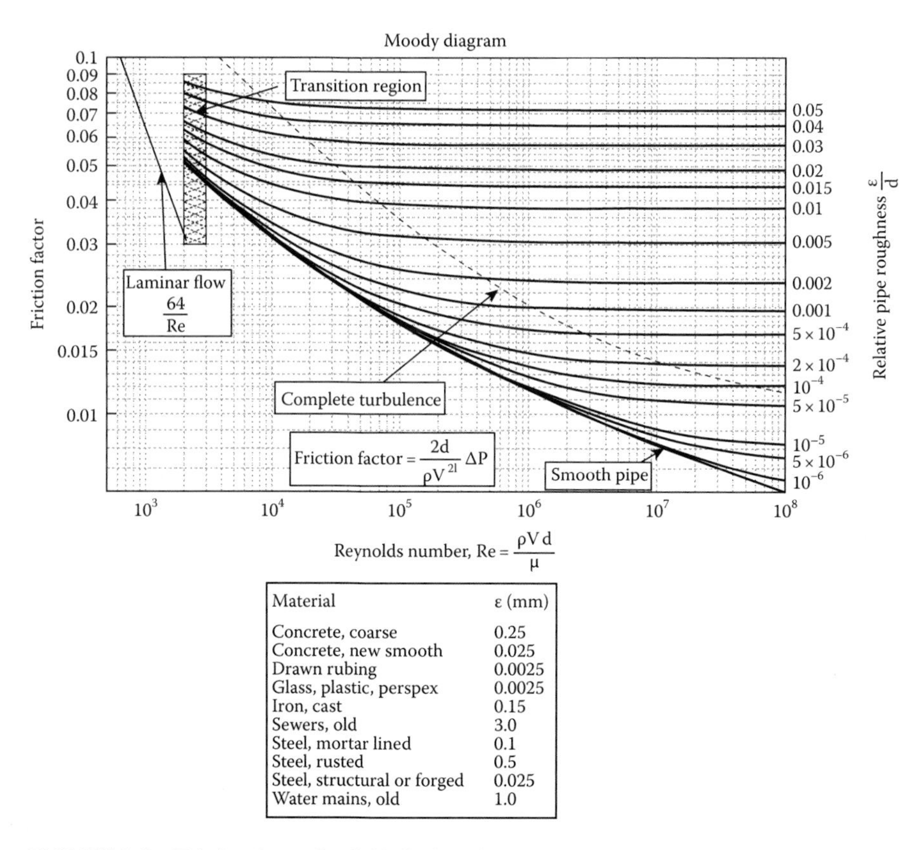

Material	ε (mm)
Concrete, coarse	0.25
Concrete, new smooth	0.025
Drawn rubing	0.0025
Glass, plastic, perspex	0.0025
Iron, cast	0.15
Sewers, old	3.0
Steel, mortar lined	0.1
Steel, rusted	0.5
Steel, structural or forged	0.025
Water mains, old	1.0

FIGURE 5.4 Friction factor for fluids inside pipes.

5.7 PIPING AND PUMPS

5.7.1 Introduction

Pipelines along with pumps are needed as an efficient means of transporting fluids, in general, such as natural gas and other important fossil fuels, speedily, safely, and smoothly. Pipelines need to be constantly and reliably operated and monitored in order to ensure maximum operating efficiency, safe transportation, and minimal downtimes and to maintain environmental and quality standards. Powerful pumps, on the other hand, are needed for transport of fluids, to overcome frictional losses and changes in elevation.

In general, two types of fluids are encountered: noncompressible fluids and compressible fluids.

5.7.2 Piping

5.7.2.1 Sizing of Pipelines

Choosing a pipeline size and determining wall thickness represent a major target in the design of piping system. In this respect, one should be aware of two fundamental concepts

1. The diameter of a pipe is a function of the flow rate of the fluid: D = f(Q)
2. The thickness of a pipe is a function of the working pressure inside the pipe: t = f(p)

By sizing, we mean to determine the pipe diameter first. An engineer in charge must specify the diameter of the pipe that will be used in a given piping system. Normally, the economic factor must be considered in determining the optimum pipe diameter.

To calculate the pipe diameter for noncompressible fluids, one can apply the following well-known equation:

$$\begin{aligned} Q &= u \cdot A(\text{cross}-\text{sectional area of pipe}) \\ &= u[\pi/4]d^2 \end{aligned}$$

Pipe diameter, d, is readily calculated from this equation for a specified flow rate Q (bbl/h) and for an assumed fluid velocity u (ft/s).

5.7.2.2 Economic Balance in Piping and Optimum Pipe Diameter

When pumping of a specified quantity of fluid over a given distance, two alternatives exist and a decision has to be made

1. Whether to use a large-diameter pipe with a small pressure drop
2. Whether to use a smaller-diameter pipe with a greater pressure drop

The first alternative involves a higher capital cost with lower running costs; the second involves a lower capital cost with higher running costs specifically because of the need for more pumps.

So it is necessary to arrive at an economic balance between the two alternatives. Unfortunately, there are no hard and fast rules or formulas to use; every case is different.

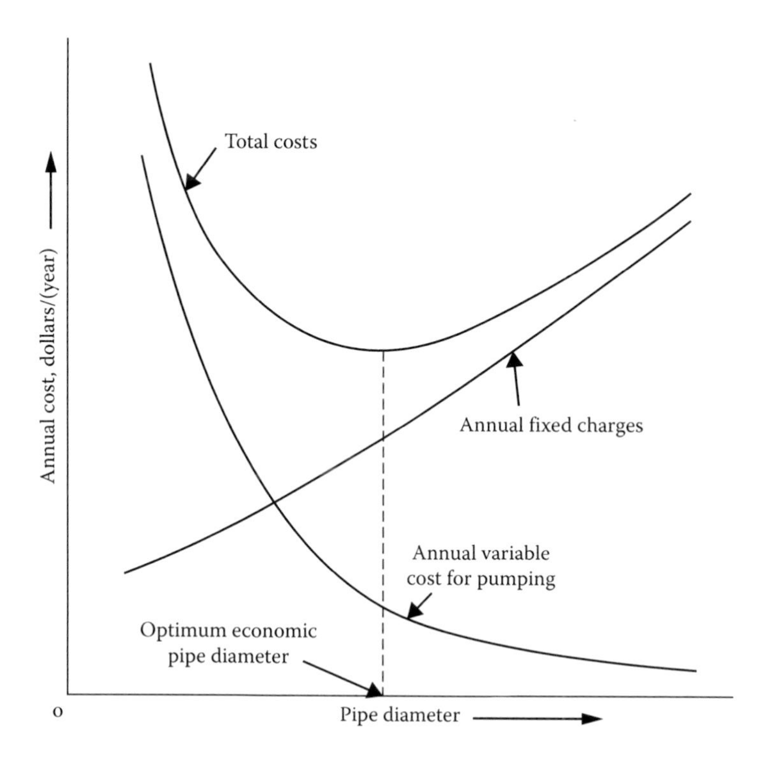

FIGURE 5.5 Optimum economic pipe diameter.

The costs of actual pumping equipment undoubtedly must be considered, but the area in which the pipes will *run* is also important. For instance, to obtain the same pumping effort in the desert as opposed to a populated area could involve much higher costs in the form of providing outside services and even creating a small, self-contained township.

In the flow of oil in pipes, the fixed charges are the cost of the pipe, all fittings, and installation. All these fixed costs can be related to pipe size to give an approximate mathematical expression for the sum of the fixed charges.

In the same way, direct costs, or variable costs, comprising mostly the costs of power for pressure drop plus costs of minor items such as repairs and maintenance, can be related to pipe size. For a given flow, the power cost decreases as the pipe size increases. Thus, direct costs decrease with pipe size. And total costs, which include fixed charges, reach a minimum at some optimum pipe size. The ultimate solution leading to the optimum economic diameter is found from the graph shown in Figure 5.5.

5.7.2.3 Stepwise Procedure to Calculate the Wall Thickness

- Determine D_i (as shown in 5.7.2.2), guided by the allowable pressure drop in a pipeline (ΔP).
- Select a material of construction; S (tensile strength) is determined.
- Knowing our operating pressure, Schedule No = 1000 P/S is calculated.
- If severe corrosion is anticipated in your pipe system, choose a larger schedule no.

- Pick up a nominal pipe size with the specified schedule no. that gives D_i for our flow equal or slightly greater than D_i obtained before.
- As a final check, use the following equation to calculate the safe working pressure:

  Schedule No = 1000 P/S = $2000(t_m/D_{av})$; solving this relationship to obtain

$$P_s = 2S_s\left[\frac{t_m}{D_{av}}\right]$$

where

- P is the operating pressure
- P_s is the safe working pressure
- S_s is the safe working fiber stress
- S is the tensile strength, the greatest longitudinal stress a material can bear without tearing apart
- t_m is the minimum thickness of pipe
- D_{av} is the average diameter of D_i and D_o

There are many factors that affect the pipe wall thickness, which include

- The maximum and working pressures
- Maximum and working temperatures
- Chemical properties of the fluid
- The fluid velocity
- The pipe material and grade
- The safety factor or code design application

5.7.2.4 Relationship between the Pipe Diameter and the Pressure Drops

Two scenarios could be followed: either to assume a value for the velocity u and calculate ΔP or to consider an allowable value for the pressure drop ΔP and calculate the corresponding u.

The *first scenario* is illustrated by the following block diagram:

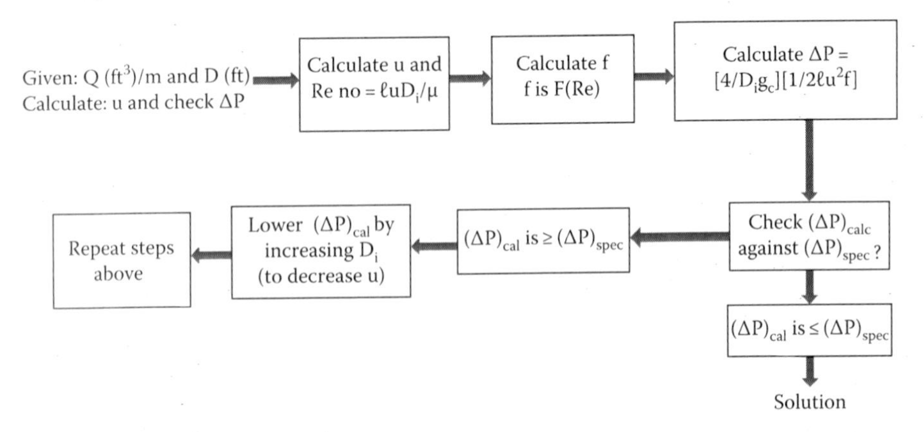

To determine the fluid velocity in a pipe, the *rule-of-thumb* economic velocity for turbulent flow is used, as given next:

Type of Fluid	Reasonable Assumed u (ft/s)
Water or fluid similar to water	3–10
Low-pressure steam (25 psig)	50–100
High-pressure steam (>100 psig)	100–200

Economies of scale is the major element in pipeline economies. From a theoretical point of view, doubling the pipeline diameter will tend to increase the amount delivered by more than fourfold in a given period of time—other factors remaining constant. This implies that total cost might double, while the cost per unit delivered would decline. In planning a pipeline system, crude oil moves at a speed of 5 km/h via pipeline, roughly a walking speed.

Pipeline capacity is normally defined as the quantity (volume) of oil and gas required to maintain a full pipeline. The static capacity of a pipeline is usually expressed as a volume per unit length (e.g., bbl/ft). On the other hand, the fluid volume passing through a pipeline in a specific time period will depend on the following factors:

- Initial pressure
- Flow characteristics and fluid density
- Ground elevation
- Delivery pressure

5.7.3 Pumps

5.7.3.1 Introduction

A fluid moves through a pipe or a conduit by increasing the pressure of the fluid using a pump, which supplies the driving force for flow. In doing so, power must be provided to the pump. There are six basic means that cause the transfer of fluid flow: gravity, displacement, centrifugal force, electromagnetic force, transfer of momentum, and mechanical impulse. Excluding gravity, *centrifugal force* is the means most commonly used today in pumping fluids.

Centrifugal force is applied by using *centrifugal pump or compressor*, where the basic function of each is the same. That is to produce KE by the action of centrifugal force and then converting this KE into potential energy (PE) by efficient reduction of the velocity of the flowing fluid.

Fluid flow in pipes applying centrifugal devices has in general the following basic advantages and features:

- Fluid discharge is relatively free from pulsations.
- There is no limitation on throughput capacity of the operating pump.
- Discharge pressure is a function of the fluid density, that is, $P = f(\ell_f)$.
- Efficient performance is provided in a simple way with low first cost.

5.7.3.2 Classification and Types of Pumps

1. Pumps can be classified into three major groups according to the method they use to move the fluid:
   a. Direct lift
   b. Displacement
   c. Gravity pumps
2. Pumps can also be classified by their method of displacement as
   a. Positive displacement pumps
   b. Impulse pumps
   c. Velocity pumps
   d. Gravity pumps
3. Pumps operate by some mechanism
   a. Reciprocating
   b. Rotary
4. Mechanical pumps may be
   a. Submerged in the fluid they are pumping
   b. Placed external to the fluid

A concise summary for the comparison between different types of pump is found in Table 5.1.

Pumps are used for many different applications. Understanding which pump type one needs for this application is very important. For the oil and gas industry, some basic features are listed next:

1. Pumps should handle the fluids with low shear and least damage to droplet sizes causing no emulsions for the effective separation of water from oil.
2. Pumps should be self-priming and experience no gas locking.
3. The requirement of having low net positive suction head is an advantage. This is advantageous for vessel-emptying applications such as closed-drain drums or flare knockout drums or any applications encountering high-vapor pressure liquids.
4. Pumps should handle multiphase fluids.

TABLE 5.1
Comparison between Types of Pumps

Type of Pump	Features
Centrifugal	Most common; high capacity; discharge lines can be shut off (safe). Handle liquids with solids.
Reciprocating	Low capacity and high head; can handle viscous fluids; used to discharge bitumen (asphalt) in vacuum distillation columns.
Rotary positive displacement	Combination of rotary motion and positive displacement; used in gas pumps, screw pumps, and metering pumps.
Air displacement	Nonmechanical, airlift type; used for *acid eggs* and jet pumps

5.7.3.3 How to Select a Pump?

The following guide to pump types should prove to be helpful for better understanding the advantages and specifications for each pump type (Hydraulic Institute, n.d., FAO Corporation Document Repository, 2011):

- *American Petroleum Institute (API) process pumps*: Designed to meet the 610 standard set by the API.
- *Boiler feed pumps*: Built to control the amount of water that enters a boiler. They are centrifugal pumps and most are multistage.
- *Chemical pumps*: Built to handle abrasive and corrosive industrial materials. They can be of either centrifugal or positive displacement type.
- *Circulator pumps*: Used to circulate fluid through a closed or looped system. They are usually centrifugal pumps, but a few use positive displacement technologies.
- *Dewatering pumps*: A dewatering process involves using a centrifugal pump (submersible or vertical turbine) to remove water from a construction site, pond, mine shaft, or any other area.
- *Fire pumps*: A type of centrifugal pump used for firefighting. They are generally horizontal split case, end suction, or vertical turbine.
- *High-pressure pumps*: Used in many applications including water blast, hydromining, and jet cutting. They can be a wide variety of pump types including positive displacement pumps, rotary pumps, and reciprocating pumps or centrifugal pumps.
- *Industrial pumps*: Used in industrial applications such as slurry, wastewater, industrial chemicals, oil, and gas. There are dozens of different industrial pumps both in positive displacement and centrifugal pump types.
- *Marine pumps*: Built to pump seawater. They are often used in large saltwater tanks to continuously circulated water so it stays fresh.
- *Mixed flow pumps*: Incorporate the features of both axial flow pumps and radial flow pumps. Axial flow pumps operate on a vertical plane and radial flow pumps operate on a horizontal plane to the flow direction of water.
- *Mud pumps*: Built to transfer heavy sludge or mud. Some larger versions are used to pressure. They are sometimes used on oil rigs to pressurize and circulate fluid.
- *Petrochemical pumps*: Made to transfer petroleum products that are often very viscous and corrosive. They can be magnetic drive pumps, diaphragm pumps, piston pumps, and others.
- *Pneumatic pumps*: Use compressed air to pressurize liquid through the piping system.
- *Pressure pumps*: Used to create either high or low pressure. They can be metering pumps and sometimes booster pumps.
- *Process pumps*: Are many times centrifugal pumps or positive displacement pumps used in process applications. The type of pump and construction details varies depending on the application in which these pumps are used.
- *Slurry pumps*: A heavy-duty pump that is made to handle thick, abrasive slurries. They are made of durable materials and capable of handling abrasive fluids for long periods of time.

- *Solar pumps*: Powered by the sun. They can be positive displacement pumps or centrifugal pumps.
- *Water pumps*: A type of equipment used to move water through a piping system. They rely upon principles of displacement, gravity, suction, and vacuums to move water. They can be either positive displacement or centrifugal pumps.
- *Well pumps*: Designed to draw water to the surface from an underground water source. Depending on the well depth and configuration, these pumps can be jet pumps, centrifugal pumps, or submersible pumps.

In conclusion, the final selection of a pump for a particular operation is influenced by many factors, including the following:

- Pump capacity (size) that is a function of the flow rate to be pumped
- Fluid properties, both physical and chemical
- Operating conditions
- Type of power supply
- Type of flow distribution

5.7.3.4 Calculation of the Horsepower for a Pump

This method is recommended to calculate the horsepower (HP) for a pump as a function of the flow rate and the total equivalent head or the gauge pressure. It is much simpler to apply than using the mechanical energy equation:

$$\text{HP (hydraulic horsepower)} = \frac{[H \cdot \ell \cdot Q]}{3960} = \frac{[P \cdot Q]}{1714}$$

$$\text{Brake HP (actual)} = \frac{\text{HP(hydraulic)}}{\alpha}$$

where

H is the head in ft

ℓ is the specific gravity

Q is the flow rate (gpm)

P is the gauge pressure in $lb/in.^2$

α is the pump efficiency; 60% is used for centrifugal pumps

END-OF-CHAPTER SOLVED EXAMPLES

Example 5.1

Assume frictionless flow in a long, horizontal, conical pipe. The diameter is 2.0 ft at one end and 4.0 ft at the other. The pressure head at the smaller end is 16 ft of water. If water flows through this cone at a rate of 125.6 ft^3/s, find the pressure head at the larger end.

Solution

By applying Bernoulli's equation between the two ends:

$$\frac{p_1}{\rho}\frac{g_c}{g}+Z_1+\frac{u_1^2}{2g}=\frac{p_2}{\rho}\frac{g_c}{g}+Z_2+\frac{u_2^2}{2g}$$

$$Z_1 = Z_2 = 0.0 \text{ (horizontal pipe)}$$

$$V_1=\frac{V^o}{A_1}=\frac{125.6\ \text{ft}^3/\text{s}}{\pi(1)^2\ \text{ft}^2}=40\ \text{ft/s}$$

$$V_2=\frac{V^o}{A_2}=\frac{125.6\ \text{ft}^3/\text{s}}{\pi(2)^2\ \text{ft}^2}=10\ \text{ft/s}$$

Substitute in Bernoulli's equation:

$$16+0.0+\frac{(40)^2}{2\times 32.17}=\frac{p_2}{\rho}\frac{g_c}{g}+0.0+\frac{(10)^2}{2\times 32.17}$$

$$\therefore \text{Outlet pressure head}=\frac{p_2}{\rho}\frac{g_c}{g}=39.9\ \text{ft}$$

Example 5.2

Water is pumped from a large reservoir to a point 65 ft higher than the reservoir. How many feet of head must be added by the pump if 8000 lb_m/h flows through a 6 in. pipe and the frictional head loss is 2 ft? The density of the fluid is 62.4 lb_m/ft^3 and the pump efficiency is 60%. Assume the kinetic energy correction factor equals 1.

Solution

$$\frac{p_1}{\rho}+Z_1\frac{g}{g_c}+\frac{V_1^2}{2\alpha_1 g_c}+\eta W=\frac{p_2}{\rho}+Z_2\frac{g}{g_c}+\frac{V_2^2}{2\alpha_2 g_c}+\sum F$$

Multiply this equation by the factor g_c/g to be in terms of head:

$$\frac{p_1}{\rho}\frac{g_c}{g}+Z_1+\frac{V_1^2}{2\alpha_1 g}+\eta H_p=\frac{p_2}{\rho}\frac{g_c}{g}+Z_2+\frac{V_2^2}{2\alpha_2 g}+H_f$$

where

- H_p is the pump head
- H_f is the frictional head loss

$$\eta H_p = (Z_2 - Z_1) + \frac{(V_2^2 - V_1^2)}{2g} + \frac{(p_2 - p_1)}{\rho}\frac{g_c}{g} + H_f$$

To use the modified form of Bernoulli's equation, reference points are chosen at the surface of the reservoir (point 1) and at the outlet of the pipe (point 2). The pressure at the surface of the reservoir is the same as the pressure at the exit of the pipe, that is, atmospheric pressure. The velocity at point 1 will be essentially zero.

Manual solution is presented first, followed by solving the problem by Excel.

$$V_2 = \frac{V^o}{A_2} = \frac{m^o/\rho}{\pi(r)^2\ ft^2} = \frac{8000/(62.4 \times 3600)}{\pi(0.25)^2\ ft^2} = 0.178\ ft/s$$

$$0.6 \times H_p = 65 + \frac{((0.178)^2 - 0.0)}{2 \times 32.17} + 0.0 + 2$$

$$\eta H_p = (Z_2 - Z_1) + \frac{(V_2^2 - V_1^2)}{2g} + \frac{(p_2 - p_1)}{\rho}\frac{g_c}{g} + H_f$$

$$H_p = 111.66\ ft$$

A	B	C	D	E
		η	0.6	
		point 1	65	
		point 2	0	
		flow (m^o)	8000	
		pipe inch	6	
	frictional h	H_f	2	
		density ρ	62.4	
		pump efficiency	60%	
		K.E. factor	1	
		g_c	1	
		g	32.17	
		P_1	1	
		P_2	1	
		V_1	0	
		V_1^2	0	
		V^o	128.20513	

A	B	C	D	E	F	G	H	I	J
		A_2			(Z_2-Z_1)	65			
		r	0.25		$(V2^2-V1^2)/2g$	0.00051			
		r^2	0.0625		$(P_2-P_1)g_c/\rho*g$	0			
		time s	3600		H_f	2			
		π	3.14		η	0.6			
		$V_2 = V^o/A_2$	0.1814651	ft/s					
		V_2^2	0.0329296						
					H_p	111.668	ft		

Example 5.3

Calculate the diameter of a pipeline handling 10,000 bbl of oil per hour, assuming that the velocity of flow is about 5 ft/s.

Solution

$$Q = u(\pi/4)d^2$$

$$[10,000\ \text{bbl}/(4.2\ \text{ft}^3/\text{bbl})]/3,600\ \text{s/h} = 5\ \text{ft/s}\ [(3.1416/4)d^2]\ \text{ft}^2$$

Solving for d = 1.724 ft

Example 5.4

Find the HP for a pump that is handling 500 gpm of oil against a 1000 ft pipeline with 19 ft equivalent to pipe fittings and valves. Assume friction losses account to 20% of the total head and the specific gravity of the oil is 0.8.

Solution

Total equivalent H = 1000 + 19 + 0.2(1019) = 1223 ft

$$\text{HP} = [(1223)(500)(0.8)]/3960$$

$$= 123.5$$

$$\text{Brake HB} = 123.5/0.6$$

$$= 206$$

Example 5.5

1. Manual solution
   The pressures at two sections of a horizontal pipe, Sections 1 and 2, are 0.3 and 0.6 kgf/cm², and the diameters are 7.5 and 15 cm, respectively. Determine the direction of flow if water flows at a rate of 8.5 kg/s. State your assumptions.

   Data:

Section 1	Section 2
$P_1 = 0.3$ kgf/cm²	$P_2 = 0.6$ kgf/cm²
$D_1 = 7.5$ cm	$D_2 = 15$ cm

Mass flow rate = 8.5 kg/s

Equations

Equation of continuity:

$$\rho_1 A_1 v_1 = \rho_2 A_2 v_2$$

Bernoulli's equation
For the flow direction from 1 to 2,

$$\frac{p_1}{\rho_1 g}+\frac{v_1^2}{2g}+z_1=\frac{p_2}{\rho_2 g}+\frac{v_2^2}{2g}+z_2+h+w-q$$

Mass flow rate = volumetric flow rate x density

Solution

Volumetric flow rate = 8.5/1000 = 8.5×10^{-3} m^3/s

Section 1	Section 2
$V_1 = 8.5\times10^{-3}/(\pi D_1^2/4) = 1.924$ m/s	$V_1 = 8.5\times10^{-3}/(\pi D_2^2/4) = 0.481$ m/s
$P_1 = 0.3$ kgf/cm² = 2.9436×10^4 N/m²	$P_2 = 0.6$ kgf/cm² = 5.8872×10^4 N/m²

Assuming the flow direction is from Section 1 to Section 2:

$$29.436\times10^4/1000+1.924^2/2=5.8872\times10^4/1000+0.481^2/2+h+w-q$$

$$29.436+1.851=58.872+0.116+h+w-q$$

In our problem the work done by fluid (w) and pump work on fluid (q) are *zero.*

So to balance this equation, the quantity h has to have negative values. This is not possible. This equation will be a correct one if the flow is from *2 to 1.*

$$\text{That is, } 58.872+0.116=29.436+1.851+h$$

Therefore, the flow direction is from the end at which pressure is 0.6 kgf/cm² and diameter is 15 mm to the end at which pressure is 0.3 kgf/cm² and diameter is 7.5 mm.

2. Excel solution

Given	
$P_1 = 0.3$ kgf/cm²	$P_1 = 29{,}420.4178$ N/m²
$P_2 = 0.6$ kgf/cm²	$P_2 = 58{,}840.8355$ N/m²
$D_1 = 7.5$ cm	$V_1 = 0.0001925$ m/s
$D_2 = 15$ cm	$V_2 = 0.0000481$ m/s
Water flow rate = 8.5 kg/s	Volumetric flow rate = 0.0085 m³/s

$$\frac{p_1}{\rho_1 g}+\frac{v_1^2}{2g}+z_1=\frac{p_2}{\rho_2 g}+\frac{v_2^2}{2g}+z_2+h+W-q$$

$h + w - q = -29.42041775253$
$w = 0$
$q = 0$
$h = -29.420$

The previous equation will be a correct one if the flow is from *2 to 1*.

Therefore the flow direction is from the end at which pressure is 0.6 kgf/cm^2 and diameter is 15 mm to the end at which pressure is 0.3 kgf/cm^2 and diameter is 7.5 mm.

Example 5.6

A simple force balance on a spherical particle reaching terminal velocity in a fluid is given by

$$V_t = ((4g(\rho_p - \rho)D_p)/(3C_D\rho))\wedge 5$$

where

- V_t is the terminal velocity in m/s
- g is the acceleration due to gravity
- p_p is the particle density
- D_p is the diameter of the spherical particle in m
- C_D is the dimensionless drag coefficient

The drag coefficient on a spherical particle at terminal velocity varies with Re number as the follows:

$C_D = 24/Re$	For $Re < 0.1$
$C_D = 24 * (1 + 014 * Re\wedge 7)/Re$	For $0.1 \le Re \le 1{,}000$
$C_D = 0.44$	For $1{,}000 < Re \le 350{,}000$
$C_D = 0.19 - 8 * 10\wedge 4/Re$	For $350{,}000 < Re$

where

$Re = (D_p V_t p)/\mu$
$g = 9.80665\ m/s^2$
$\rho = 994.6\ kg/m^3$
$p_p = 1800\ kg/m^3$
$\mu = 8.931 \times 10^{-4}\ kg/m \cdot s$
$D_p = 0.000208$

Calculate the terminal velocity of spherical particle.

Solution

This problem cannot be solved without using a trial and error procedure. Therefore, you must assume a velocity to calculate the C_D that is important to calculate a new velocity. Write the following code:

```
g = 9.80665; p = 994.6; pp = 1800; mu = 8.931e-4; Dp =
0.000208;
% Assume vt of any initial terminal velocity for first
trail vt = 1;
```

```
Velocity(1) = vt;
for m = 1:1:20
Re = (Dp*vt*p)/mu;
if Re < 0.1
CD = 24/Re;
elseif Re ≥ 0.1 & Re ≤ 1000
C_D = 24*(1 + 014*Re^7)/Re;
elseif Re>1000 & Re ≤ 350,000
CD = 0.44;
elseif Re>350,000
CD = 0.19-8*10^4/Re;
end
vt = ((4*g*(pp-p)*Dp)/(3*CD*p))^5;
Velocity(m + 1) = vt;
if abs (Velocity(m + 1)-Velocity(m))<.0001
break end
end
Velocity'
```

This code gives the following result

```
ans =
  1.0000
  0.0538
  0.0251
  0.0190
  0.0170
  0.0163
```

Example 5.7

Calculate the minimum area and diameter of a thickener with a circular basin to treat

Solid Concentration (kg/m³)	Settling Velocity (μm/s)
100	148
200	91
300	55.33
400	33.25
500	21.4
600	14.5
700	10.29
800	7.33
900	5.56
1000	4.2
1100	3.27

0.1 m^3/s of a slurry of a solid concentration of 150 kg/m^3. The results of batch settling tests are as follows:

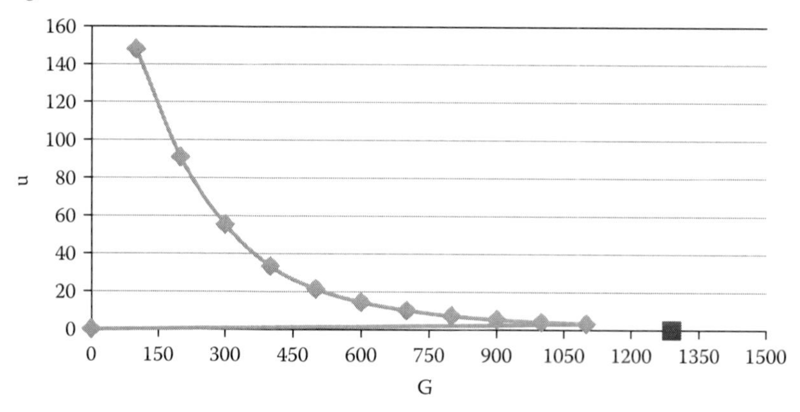

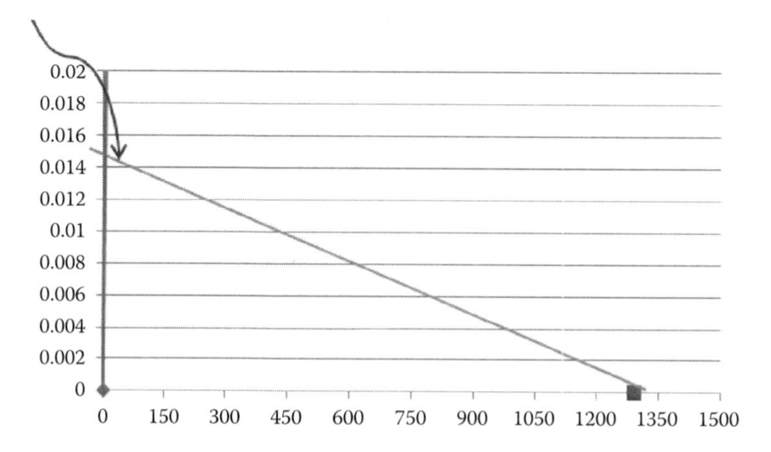

A value of 1290 kg/m^3 for underflow concentration was selected from a retention time test. Estimate the area and the diameter of the thickener and the underflow volumetric flow rate assuming total separation of all solids and clear overflow is obtained.

Solution

The settling rate of the solids, G kg/m^2s, is calculated as $G = u_s c$, where u_s is the settling velocity (m/s) and c the concentration of solids (kg/m^3), and the data are plotted as shown in the diagram. From the point u = 0 and c = 1290 kg/m^3, a line is drawn that is tangential to the curve. This intercepts the axis at G.

$$G = 0.0154\ kg/m^2s$$

The area of the thickener is then

$$A = (0.1 \times 150)0.0154 = 974\ m^2$$

and the diameter is

$$d = [(4 \times 974)/\pi]^{\wedge}05 = 352\ \text{m}$$

The volumetric flow rate of underflow, obtained from a mass balance, is

$$= [(0.1 \times 150)/1290] = 0.0116\ \text{m}^3/\text{s}$$

REFERENCES

FAO Corporation Document Repository, Water lifting devices, pump classifications. Fao.org., http://www.fao.org/docrep/010/ah810e/AH810E05.htm\15.3.1, retrieved on May 25, 2011.

Hydraulic Institute, http://www.pumps.org/, retrieved on May 25, 2011.

6 Heat Transmission

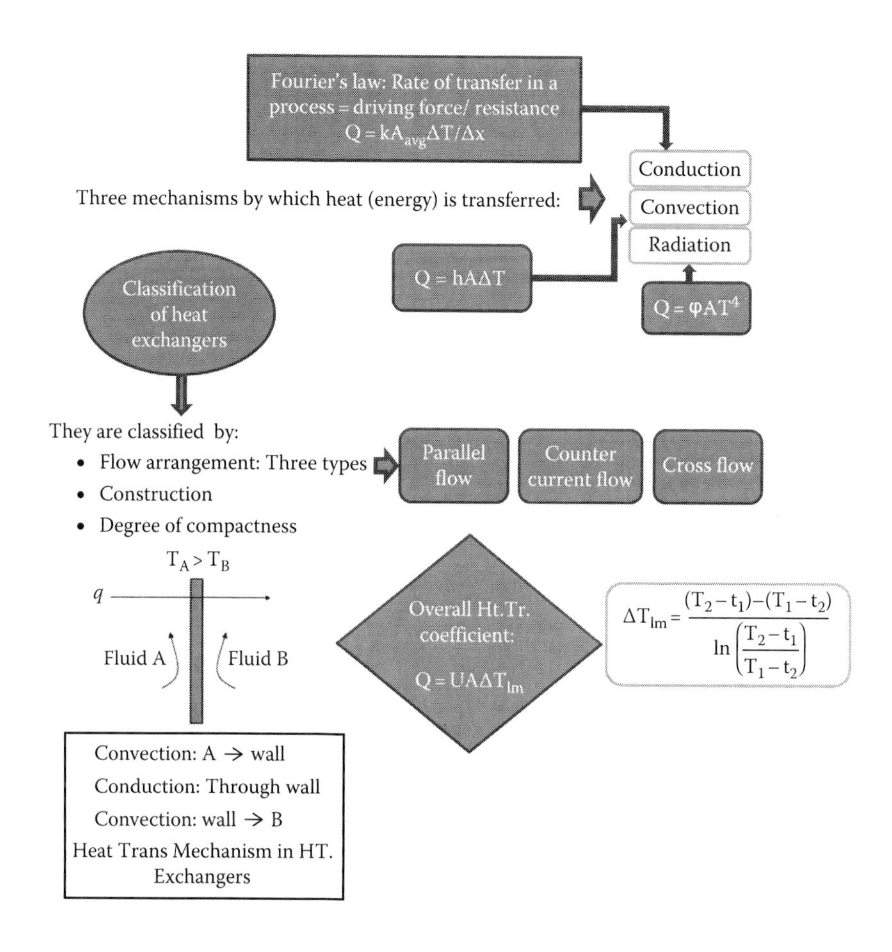

6.1 INTRODUCTION

Heat transfer, as a topic, is concerned with two main items:

1. Temperature
2. Flow of heat

Whenever a temperature difference exists in a medium or between media, heat transfer must occur. Temperature represents the amount of thermal energy available, whereas heat flow represents the movement of thermal energy from place to place.

Heat transfer always occurs from a region of high temperature to another region of lower temperature. Heat transfer changes the internal energy of both systems involved according to the first law of thermodynamics. Heat transfer is a process function (or path function), as opposed to functions of state; therefore, the amount of heat transferred in a thermodynamic process that changes the state of a system depends on how that process occurs and not only on the net difference between the initial and final states of the process.

6.2 MODES OF HEAT TRANSFER

The fundamental modes of heat transfer are as given.

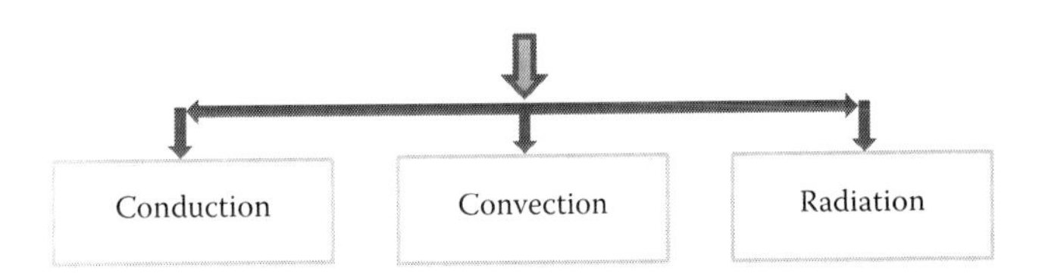

In many practical problems, these three mechanisms combine to generate the total energy flow.

1. *Conduction*: The transfer of energy between objects that are in physical contact. Thermal conductivity is the property of a material to conduct heat. It is evaluated primarily in terms of Fourier's law for heat conduction. Regions with greater molecular kinetic energy will pass their thermal energy to regions with less molecular energy through direct molecular collisions, a process known as conduction. In metals, a significant portion of the transported thermal energy is also carried by conduction-band electrons. Conduction happens in solids and fluids (direct contact between neighboring atoms or molecules through intervening matter without bulk motion of the matter).
2. *Convection*: The transfer of energy between an object and its environment, due to fluid motion. The average temperature is a reference for evaluating properties related to convective heat transfer.

   Convective heat transfer, or convection, is the transfer of heat from one place to another by the movement of fluids, a process that is essentially the transfer of heat via mass transfer. Bulk motion of fluid enhances heat transfer in many physical situations, such as between a solid surface and the fluid. Convection is usually the dominant form of heat transfer in liquids and gases. Although sometimes discussed as a third method of heat transfer, convection is usually used to describe the combined effects of heat conduction within the fluid (diffusion) and heat transference by bulk fluid flow streaming.

3. *Radiation*: The transfer of energy from the movement of charged particles within atoms is converted to electromagnetic radiation. Thermal radiation occurs through a vacuum or any transparent medium (solid or fluid). It is the transfer of energy by means of photons in electromagnetic waves governed by the same laws. Thermal radiation is energy emitted by matter as electromagnetic waves, due to the pool of thermal energy in all matter with a temperature above absolute zero. Thermal radiation propagates without the presence of matter through the vacuum of space.

6.3 HEAT TRANSFER BY CONDUCTION

6.3.1 Fourier's Law

In all transport processes, such as momentum, heat, and mass transfer, the basic rate of transfer is determined by the following formula:

Rate of transfer in a process = Driving force/resistance

Fourier's law is the fundamental differential equation for heat transfer by conduction:

$$\frac{dq}{d\theta} = Q = -kA\frac{dT}{dx} \tag{6.1}$$

where

$dq/d\theta$ is the rate of flow of heat
Q is the quantity per unit time
A is the area at right angles to the direction of heat flow
$-dT/dx$ is the rate change of temperature with a distance (temperature gradient)
k is the thermal conductivity, which is a function of the material through which heat flows

The equation for the case of *one-dimensional steady-state heat conduction*, Figure 6.1, is obtained as follows:

1. Rearranging Equation 6.1:

$$\frac{Qdx}{A} = -kdT \tag{6.2}$$

2. Integrating Equation 6.2 between the limits $x_1 - x_2$ and $T_1 - T_2$:

$$Q = kA_{avg}\Delta T / \Delta x \tag{6.3}$$

There are several ways to correlate the geometry, physical properties, and temperature difference of an object with the rate of heat transfer through the object. In heat transfer by conduction, the most common means of correlation is through Fourier's law of conduction. The law, stated by the equation, is applicable most often in objects

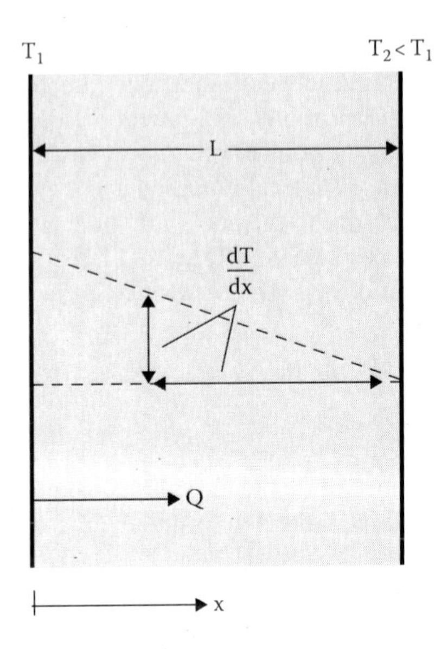

FIGURE 6.1 1D conduction.

of rectangular or cylindrical form (e.g., pipes and cylinders). Equation 6.3 is used for the case of those of rectangular form, while Equation 6.4 is designated for those of cylindrical form:

$$Q = -kA\left(\frac{\Delta T}{\Delta r}\right) \tag{6.4}$$

where
- Q is the rate of heat transfer (Btu/h)
- Q/A is the heat flux ($Btu/h \cdot ft^2$)
- A is the cross-sectional area of heat transfer (ft^2)
- x is the thickness of slab (ft)
- r is the thickness of cylindrical wall (ft)
- T is the temperature difference (°F)
- k is the thermal conductivity of slab (Btu/h · ft · °F)

6.3.2 Thermal Conductivity

The heat transfer characteristics of a solid material are measured by a property called the *thermal conductivity* (k) measured in Btu/h · ft · °F. It is a measure of a substance's ability to transfer heat through a solid by conduction. The thermal conductivity of most liquids and solids varies with temperature. For vapors, it depends upon pressure. Thermal conductivity varies with temperature but not always in the same direction.

6.3.3 Conduction through a Hollow Cylinder

In many applications in the process industry, the need arises to calculate the heat transfer through a thick-walled cylinder, as in pipes. Consider a hollow cylinder with an inside radius r_1, where the temperature is T_1, and an outside radius r_2, where the temperature is T_2, having a length L measured in meters. Heat is flowing radially from the inside surface to the outside. Equation 6.4, as given earlier, is readily applied, with distance dr instead of dx.

It can be shown that the final equation to be used is

$$Q = kA_{lm}\frac{T_1 - T_2}{r_2 - r_2}$$

where A_{lm} refers to the log mean area = $(A_2 - A_1)/\ln(A_1/A_2)$.

6.3.4 Conduction through Solids in Series

In the case of a multilayer wall of more than one material present, for example, with three layers, having resistances R_A, R_B, and R_C and a driving force of $T_1 - T_4$, the final equation to be used is represented as

$$Q = [(\Delta T): T_1 - T_4]/\text{sum of the individual resistances of each wall}$$

$$Q = [T_1 - T_4]/\{R_A + R_B + R_C\}$$

6.3.5 Solved Examples

Example 6.1

The heat transfer rate through a section of insulating material (as shown in Figure 6.2) that measures 1 ft^2 in cross-sectional area is equal to 1000 Btu/h. The thickness is 1 in. and the thermal conductivity is 0.12 Btu/h·ft·°F.

(a) Compute the temperature difference across the material.
(b) Show the temperature profile along the distance by using MATLAB®, assuming the initial temperature T_o = 400°F.

Solution

(a) $\dot{Q} = kA\left(\frac{\Delta T}{\Delta X}\right)$

Solving for ΔT,

$$\Delta T = \dot{Q}\left(\frac{\Delta X}{kA}\right)$$

$$\Delta T = \frac{1000\ (\text{Btu/h})(1/12\ \text{ft})}{0.12\ \text{Btu/h}\cdot\text{ft}\cdot{}^\circ\text{F}}$$

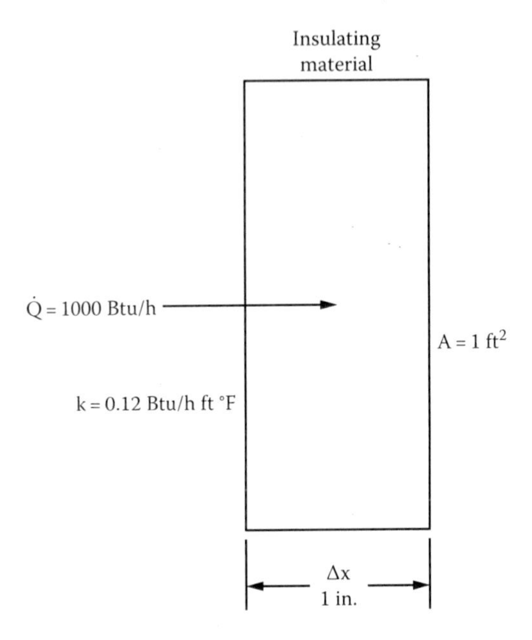

FIGURE 6.2 Conduction through a slab.

(b) Calculations by MATLAB (see Figure 6.3)

```
>> x = [0 : 0.01 : 1/12]

>> q = 1000

>> k = 0.12

>> a = 1

>> temperature = 800 - [q.*x]/[k*a]

temperature = 1.0e + 002*
```

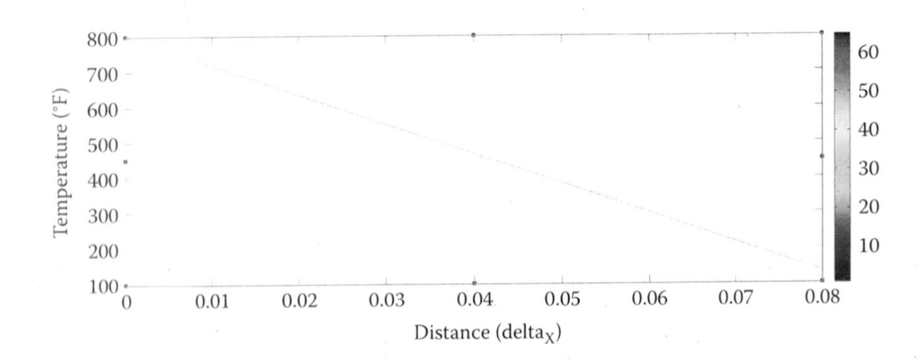

FIGURE 6.3 Distance versus temperature at steady state.

Columns 1 through 6

```
8.000000000000000    7.166666666666666    6.333333333333332
5.500000000000000    4.666666666666666    3.833333333333333
```

Columns 7 through 9

```
3.000000000000000    2.166666666666666    1.333333333333333

>> plot(x,temperature)
>
```

Example 6.2

A concrete floor with a conductivity of 0.8 Btu/h·ft·°F measures 30 ft by 40 ft with a thickness of 4 in. The floor has a surface temperature of 70°F and the temperature beneath it is 60°F. What are the heat flux and the heat transfer rate through the floor?

Solution

A direct application of $Q = q/A = -k\{\Delta T/\Delta x\}$

$$= 0.8(10/0.333)$$

$$= 24 \text{ Btu/h} \cdot \text{ft}^2$$

6.4 HEAT TRANSFER BY CONVECTION

6.4.1 Introduction

Convection involves the transfer of heat by the motion and mixing of *macroscopic* portions of a fluid (i.e., the flow of a fluid past a solid boundary). The term natural convection is used if this motion and mixing is caused by density variations resulting from temperature differences within the fluid. The term *forced convection* is used if this motion and mixing is caused by an outside force, such as a pump or fan (as shown in Figure 6.4). The transfer of heat from a hot-water radiator to a room is an example of heat transfer by *natural convection*. The transfer of heat from the surface

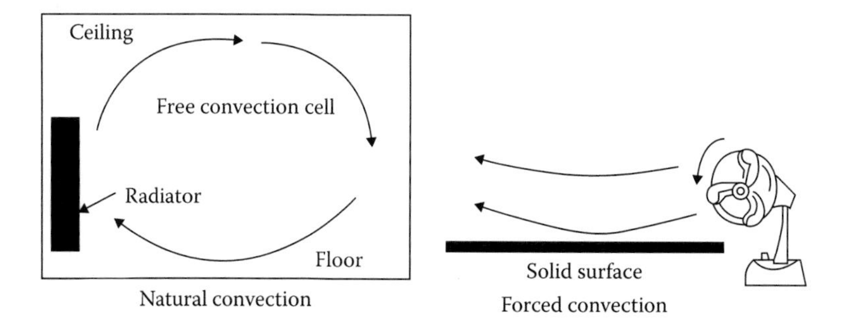

FIGURE 6.4 Schematic of natural convection versus forced convection.

of a heat exchanger to the bulk of a fluid being pumped through the heat exchanger is an example of forced convection.

Heat transfer by convection is more difficult to analyze than heat transfer by conduction because no single property of the heat transfer medium, such as thermal conductivity, can be defined to describe the mechanism. Heat transfer by convection varies from situation to situation (upon the fluid flow conditions), and it is frequently coupled with the mode of fluid flow.

In practice, the analysis of heat transfer by convection is treated empirically (by direct observation). Convection heat transfer is treated empirically because of the factors that affect the stagnant film thickness such as fluid velocity, fluid viscosity, heat flux, surface roughness, and the type of flow (single phase/double phase).

6.4.2 Methodology of Calculation

Convection involves the transfer of heat between a surface at a given temperature (T_s) and fluid at a bulk temperature (T_b). The exact definition of the bulk temperature (T_b) varies depending on the details of the situation. For flow adjacent to a hot or cold surface, T_b is the temperature of the fluid *far* from the surface. For boiling or condensation, Tb is the saturation temperature of the fluid. For flow in a pipe, T_b is the average temperature measured at a particular cross section of the pipe.

The basic relationship for heat transfer by convection has the same form as that for heat transfer by conduction:

$$\dot{Q} = hA\Delta T$$

where

$\dot{Q}$ is the rate of heat transfer (Btu/h)
h is the convective heat transfer coefficient ($Btu/ft^2 \cdot °F$)
A is the surface area for heat transfer (ft^2)
ΔT is the temperature difference (°F)

6.4.3 Convective Heat Transfer Coefficient

The convective heat transfer coefficient (h), defines, in part, the heat transfer due to convection. The *convective heat transfer coefficient* is sometimes referred to as a film coefficient and represents the thermal resistance of a relatively stagnant layer of fluid between a heat transfer surface and the fluid medium. Common units used to measure the convective heat transfer coefficient are $Btu/h \cdot ft^2 \cdot °F$.

The convective heat transfer coefficient (h) is dependent upon the physical properties of the fluid and the physical situation. Typically, the convective heat transfer coefficient for laminar flow is relatively low compared to the convective heat transfer coefficient for turbulent flow. This is due to turbulent flow having a thinner stagnant fluid film layer on the heat transfer surface. Values of h have been measured and tabulated for the commonly encountered fluids and flow situations occurring during heat transfer by convection.

Example 6.3

An uninsulated steam line, 22 ft long, crosses a room. The outer diameter of the steam line is 18 in. and the outer surface temperature is 280°F. The convective heat transfer coefficient for the air is 18 Btu/h·ft²·°F. Calculate the heat transfer rate from the pipe into the room if the room temperature is 72°F.

Solution

$$\begin{aligned}
Q &= hA\Delta T \\
&= h(2\pi rL)\Delta T \\
&= \left(18\ \text{Btu/h}\cdot\text{ft}^2\cdot{}^\circ\text{F}\right)\left(2(3.14)(0.75\,\text{ft})(22\,\text{ft})\right)(280\,^\circ\text{F} - 72\,^\circ\text{F}) \\
&= 3.88\times10^5\ \text{Btu/h}
\end{aligned}$$

6.5 OVERALL HEAT TRANSFER COEFFICIENT

6.5.1 Background

Many of the heat transfer processes encountered in nuclear facilities involve a combination of both conduction and convection. For example, heat transfer in a steam generator involves convection from the bulk of the reactor coolant to the steam generator inner tube surface, conduction through the tube wall, and convection from the outer tube surface to the secondary side fluid.

In cases of combined heat transfer for a heat exchanger, there are two values for h. There is the convective heat transfer coefficient h_i for the fluid film inside the tubes and a convective heat transfer coefficient h_0 for the fluid film outside the tubes. The thermal conductivity, k, and thickness, Δx, of the tube wall must also be accounted for. An additional term U_o, called the overall heat transfer coefficient, must be used instead. It is common practice to relate the total rate of heat transfer, $\dot{Q}$ to the cross-sectional area for heat transfer A_o and the overall heat transfer coefficient U_o. The relationship of the overall heat transfer coefficient to the individual conduction and convection terms is shown in Figure 6.5.

6.5.2 Discussion

Recalling the basic equation for the quantity of heat transfer per unit time:

$$Q = U_o A \Delta T$$

An example of U_o concept applied to cylindrical geometry is illustrated in Figure 6.6, which shows a typical combined heat transfer situation.

Using the figure representing flow in a pipe, it is found that

- Heat transfer by convection occurs between temperatures T_1 and T_2
- Heat transfer by conduction occurs between temperatures T_2 and T_3
- Heat transfer occurs by convection between temperatures T_3 and T_4

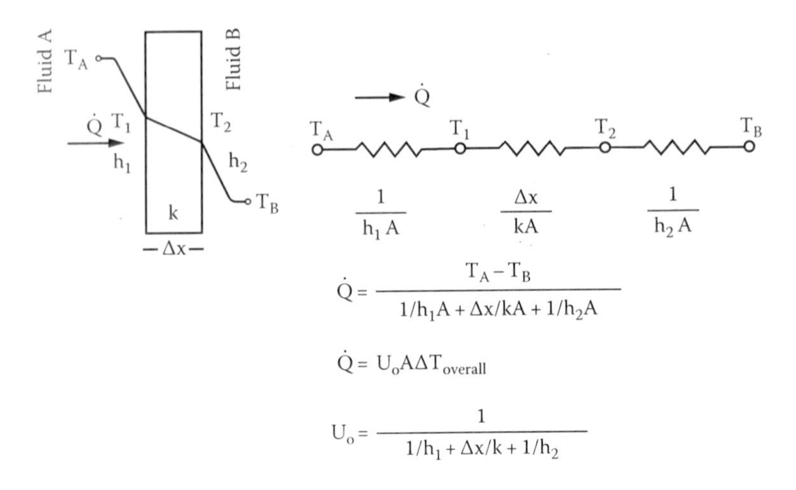

FIGURE 6.5 Relationships between U and individual conduction and convection terms.

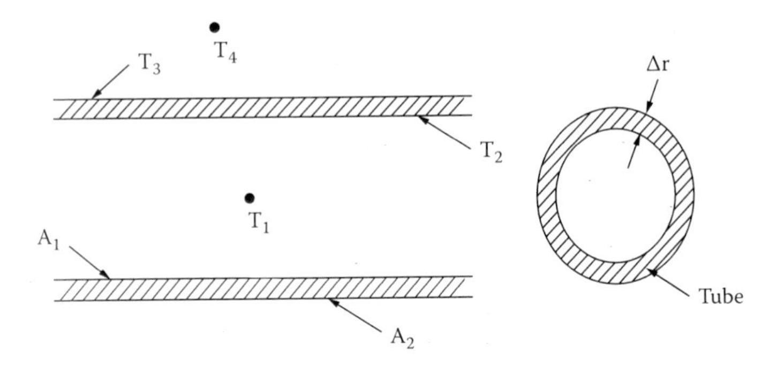

FIGURE 6.6 Example for the illustration of the concept of U_o cylindrical geometry.

Thus, there are three processes involved. Each has an associated heat transfer coefficient, cross-sectional area for heat transfer, and temperature difference. The basic relationships for these three processes can be expressed as follows:

$$\ddot{Q} = h_1A_1(T_1 - T_2)$$

$$\dot{Q} = \frac{k}{\Delta r} A_{lm}(T_2 - T_3)$$

$$\dot{Q} = h_2A_2(T_3 - T_4)$$

ΔT_o can be expressed as the sum of the three individual processes:

$$\Delta T_o = (T_1 - T_2) + (T_2 - T_3) + (T_3 - T_4)$$

If the basis relationship for each process is solved for its associated temperature difference and substituted into the expression for ΔT, the following relationship results:

$$\Delta T_o = \dot{Q}\left(\frac{1}{h_1A_1} + \frac{\Delta r}{kA_{lm}} + \frac{1}{h_2A_2}\right)$$

This relationship can be modified by selecting a reference cross-sectional area A_o:

$$\Delta T_o = \frac{\dot{Q}}{A_o}\left(\frac{A_o}{h_1A_1} + \frac{\Delta rA_o}{kA_{lm}} + \frac{A_o}{h_2A_2}\right)$$

Solving for $\dot{Q}$ results in an equation in the form $\dot{Q} = U_oA_o\Delta T_o$

$$\dot{Q} = \frac{1}{\left(\frac{A_o}{h_1A_1} + \frac{\Delta rA_o}{kA_{lm}} + \frac{A_o}{h_2A_2}\right)}A_o\Delta T_o$$

or

$$Q = UA\Delta T$$

where

$$U = \frac{1}{\left(\frac{A_o}{h_1A_1} + \frac{\Delta rA_o}{kA_{lm}} + \frac{A_o}{h_2A_2}\right)}$$

Example 6.4

Calculate the heat rate per foot of a tube from a condenser under the following conditions. T = 232°F and the outer diameter of the copper condenser tube is 0.75 in. with a wall thickness of 0.1 in. Assume the inner convective heat transfer coefficient is 2000 Btu/h·ft²·°F and the thermal conductivity of copper is 200 Btu/h·ft·°F. Also, the outer convective heat transfer coefficient is 1500 Btu/h·ft·°F.

Solution

$$U_o = \frac{1}{(1/h_1) + (\Delta r/k) + (1/h_2)}$$

$$= \frac{1}{(1/2000) + (0.1\text{ in.}/200)(1\text{ ft}/12\text{ in.}) + (1/1500)}$$

$$= 827.6\ \text{Btu/h}\cdot\text{ft}^2\cdot{}^\circ\text{F}$$

$$\dot{Q} = U_o A_o \Delta T_{lm}$$

$$\begin{aligned}\frac{\dot{Q}}{L} &= \frac{U_o A_o \Delta T_{lm}}{L} \\ &= U_o 2\pi r \Delta T_{lm} \\ &= \left(827.6\ \text{Btu/h}\cdot\text{ft}^2\cdot{}^\circ\text{F}\right)(2\pi)(0.375\ \text{in.})\left(\frac{1\ \text{ft}}{12\ \text{in.}}\right)(232^\circ\text{F}) \\ &= 37{,}700\ \text{Btu/h}\cdot\text{ft}\end{aligned}$$

6.6 HEAT TRANSFER BY RADIATION

6.6.1 Introduction

While both conductive and convective transfers involve the flow of energy through a solid or fluid substance, no medium is required to achieve radiative heat transfer. As a matter of fact, electromagnetic radiation travels more efficiently through a vacuum. The thermal radiation band is shown within the electromagnetic spectrum in Figure 6.7.

Radiant heat transfer involves the transfer of heat by electromagnetic radiation that arises due to the temperature of a body. Most energy of this type is in the infrared region of the electromagnetic spectrum although some of it is in the visible region. The term thermal radiation is frequently used to distinguish this form of electromagnetic radiation from other forms, such as radio waves, x-rays, or gamma rays. The transfer of heat from a fireplace across a room in the line of sight is an example of radiant heat transfer.

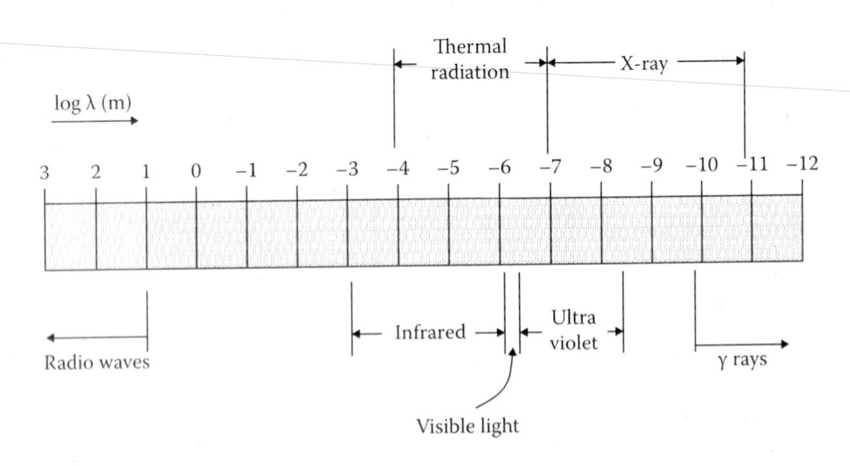

FIGURE 6.7 Band of thermal radiation within the electromagnetic spectrum.

6.6.2 Blackbody Radiation

A body that emits the maximum amount of heat for its absolute temperature is called a blackbody. Radiant heat transfer rate from a blackbody to its surroundings can be expressed by the following equation:

$$Q = \Phi A T^4 \quad (6.5)$$

where

Q is the heat transfer rate (Btu/hr)
Φ is the Stefan–Boltzmann constant (0.174 Btu/hr · ft² · °R⁴)
A is the surface area (ft_2)
T is the temperature (°R)

6.7 HEAT EXCHANGERS

6.7.1 Types of Heat Exchangers

Heat exchangers are defined as thermal devices that transfer or exchange heat from one fluid to another. They are classified either by

- Flow arrangement
- Construction
- Degree of compactness

We will elaborate on the classification by flow arrangement. Three types are known:

1. *Parallel flow*: Both fluids run side by side in the same direction.
2. *Countercurrent flow*: Similar to parallel flow, except the streams go in the opposite side to each other.
3. *Cross flow*: Here, we have the two-flow stream normal to one another.

In recuperative heat exchangers, mixing and contamination is prevented by solid walls. The dominant mechanism of heat transfer is illustrated in Figure 6.8.

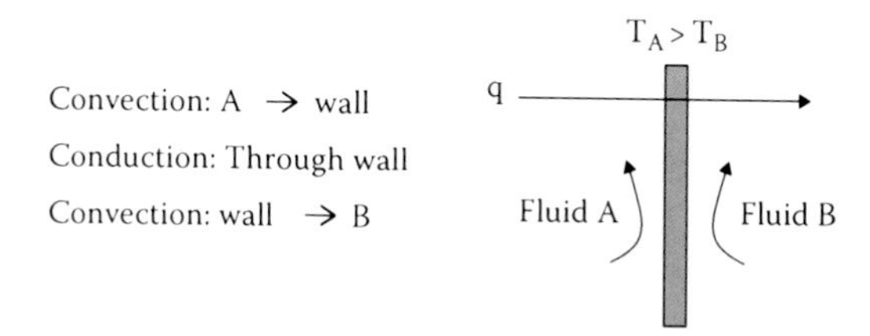

FIGURE 6.8 Heat transfer mechanism in heat exchangers.

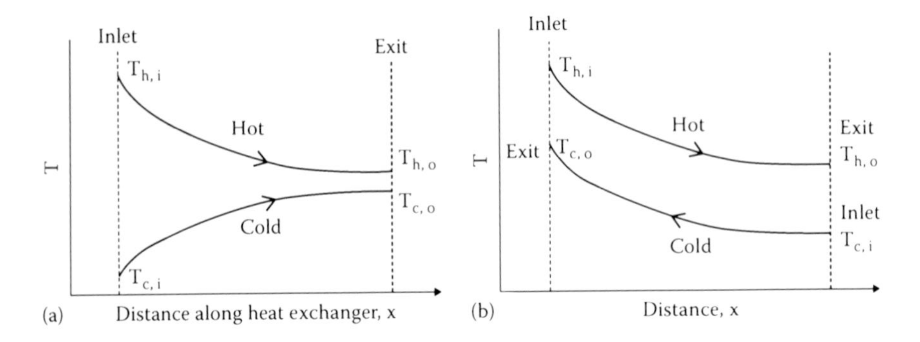

FIGURE 6.9 Temperature distribution in both flow arrangements. (a) Parallel flow and (b) counterflow.

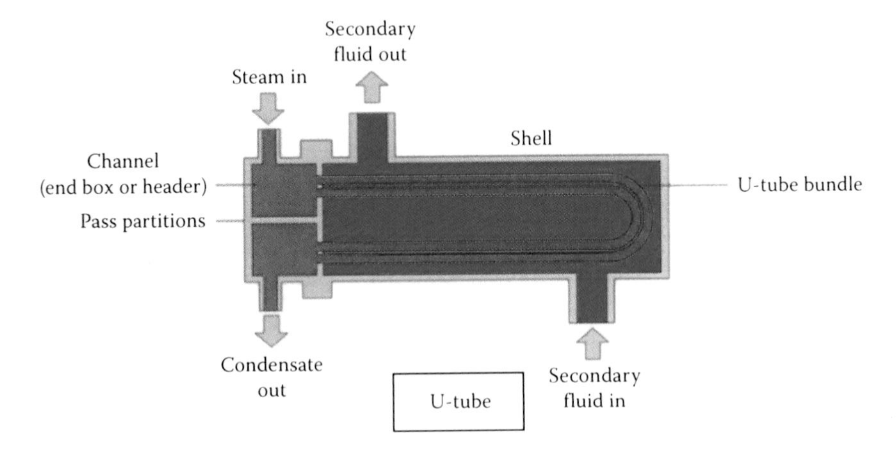

FIGURE 6.10 U-type heat exchanger.

Temperature distribution for parallel flow and counterflow is illustrated in Figure 6.9.

The two well-known types, U-type and the shell and tube heat exchangers, are diagrammatically shown in Figures 6.10 and 6.11, respectively.

6.7.2 Log Mean Temperature Difference

In heat exchanger applications, the inlet and outlet temperatures are commonly specified based on the fluid in the tubes. The temperature change that takes place across the heat exchanger from the entrance to the exit is not linear. A precise temperature change between two fluids across the heat exchanger is best represented by the *log mean temperature difference* (LMTD) or ΔT_{lm}, given by following equation:

$$\Delta T_{lm} = \frac{(T_2 - t_1) - (T_1 - t_2)}{\ln\left((T_2 - t_1)/(T_1 - t_2)\right)} \tag{6.6}$$

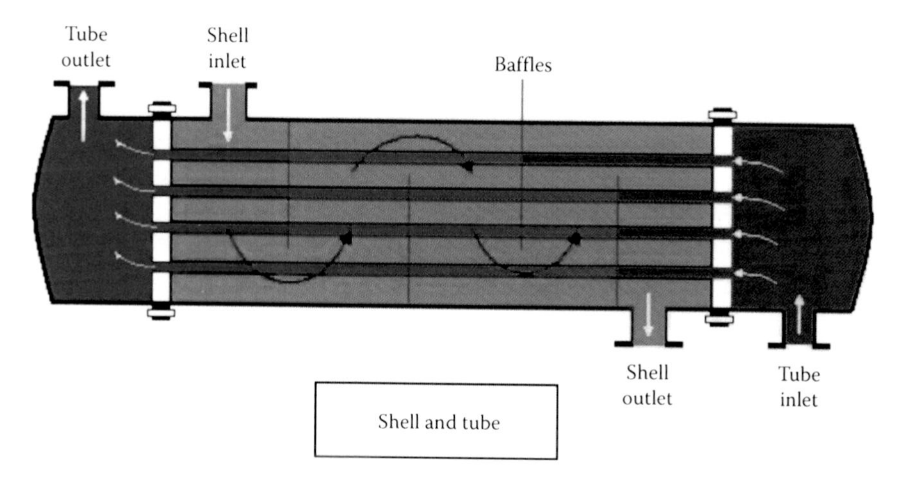

FIGURE 6.11 Shell and tube heat exchanger.

where

T_1 is the temperature of hot fluid stream at the entrance
T_2 is the temperature of hot fluid stream at the exit
t_1 is the temperature of cold fluid stream at the entrance
t_2 is the temperature of cold fluid stream at the exit

6.7.3 Design of Heat Exchangers

The simplest answer to give for what is meant by the design of a heat exchanger is to find the value of "A" using the following relationship:

$$Q = UA\Delta T$$

Given: Q, U, ΔT

Find: A

Now, it is not that simple to find the area of a heat exchanger.
The *thermal* design of a shell and tube HE is briefly described by the following procedure:

1. Consideration of process fluids in both shell and tube sides
2. Selection of the required temperature specifications
3. Limiting the shell and tube side pressure drop
4. Setting shell and tube side velocity limits
5. Finding heat transfer area including fouling factor

In addition, to carry out your design, it would require the following input:

1. Mass flow rate
2. Heat duty

3. Shell-side inlet temperature
4. Shell-side outside temperature
5. Tube-side inlet temperature
6. Tube-side outlet temperature
7. Heat transfer coefficient
8. Tube outside diameter
9. Length of tube

Example 6.5

A liquid-to-liquid countercurrent flow heat exchanger is used as part of an auxiliary system at a nuclear facility. The heat exchanger is used to heat a cold fluid from 120°F to 310°F.

Assuming that the hot fluid enters at 500°F and leaves at 400°F, calculate the LMTD for the exchanger.

Solution

$$\Delta T_2 = 400°F - 120°F = 280°F$$

$$\Delta T_1 = 500°F - 310°F = 190°F$$

$$\Delta T_{lm} = \frac{\Delta T_2 - \Delta T_1}{\ln(\Delta T_2/\Delta T_1)}$$

$$= \frac{(280°F - 190°F)}{\ln(280°F/190°F)}$$

$$= 232°F$$

END-OF-CHAPTER SOLVED EXAMPLES

Example 6.6

Consider a rectangular plate with some heat sources on each edge. Assuming these heat sources remain constant, the plate will eventually reach an equilibrium temperature. We will assume a constant heat source, while the ambient air is staying at a constant temperature. This would make one side of the cube to gain a very high temperature and the other five sides a relatively low temperature (room temperature). We are, of course, ignoring any fans or buildup of heat within the enclosure. The 2D analog to this problem would be some cross sections cut out from this cube. Consider discretizing our rectangular plate with a rectangular grid (Figure 6.12). To determine the equilibrium, we may assume that if the plate is at equilibrium and x_i is a grid point not on the boundary, then the temperature at x_i is given by the average of the temperatures of the four closest grid points to x_i. This creates a linear system for x_i, which we can then solve.

This problem corresponds to solving a partial differential equation using finite differences in two dimensions using MATLAB.

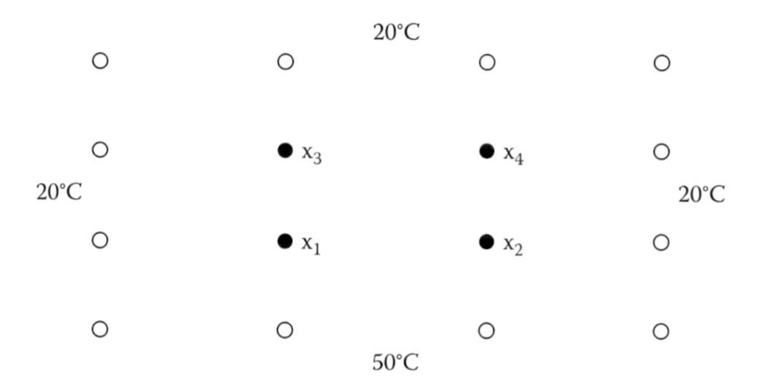

FIGURE 6.12 Rectangular plate with different heat sources on each edge.

Hint: The temperature at any point will be given by the average of the four temperatures around it.

Solution

Given:
The temperature at each point (say x_1, x_2, x_3, and x_4) is the average of the adjacent four points:

$$x_1 = \frac{50+20+x_2+x_3}{4}$$

In the same way,

$$x_2 = \frac{x_4+x_1+50+20}{4}, \quad x_3 = \frac{x_1+x_4+20+20}{4}, \quad x_4 = \frac{x_3+x_2+20+20}{4}$$

Now, make it in linear form:

$$4x_1 - x_2 - x_3 - 0x_4 = 70$$
$$-x_1 + 4x_2 + 0x_3 - x_4 = 70$$
$$-x_1 + 0x_2 + 4x_3 - x_4 = 40$$
$$0x_1 - x_2 - x_3 + 4x_4 = 40$$

This can be solved by using MATLAB:

```
A = [4  -1  -10;  - 1 4  0  -1;  -10  4  -1;  0  -1  -14]

B = [70; 70; 40; 40]

X = A\B

X = 31.2500
  = 31.2500
  = 23.7500
  = 23.7500
```

Example 6.7

Water at the rate of 30,000 lbm/h is heated from 100°F to 130°F in a shell and tube heat exchanger. On the shell side, one pass is used with water entering as the working fluid that is heated at a flow rate of 15,000 lbm/h at 200°F.

Given: $U_o = 250$ Btu/h·ft²·°F and the average water velocity in the ¾ in. 1D tubes is 1.2 ft/s.

Find: The total area, A; the number of tubes, n; and the length of the tube, L, assuming countercurrent flow and one tube pass is used

Solution (Manual)

The exit temperature of the hot water is unknown and to be calculated by

$$Q = m_c cp\Delta t_c = m_h cp\Delta t_h$$

Therefore, $\Delta t_h = 60$ and $t_{exit} = 140$°F.
Next, we calculate ΔT_{lm}. We found its value to be equal to 53.6°F.
Then, we can get the total area, A:

$$Q = m_c cp\Delta t_c = 30{,}000 \times 1 \times 30 = 9 \times 10^5 \text{ Btu/h}$$

$$= UA\Delta T_{lm}$$

$$A = \frac{9 \times 10^5}{250 \times 53.6} = 67.3 \text{ ft}^2$$

The cross sectional area of flow, S, in ft², is calculated, given velocity in ft/h and density in lbm/ft³ and mc = 30.000 lbm/h:

Therefore,

$$S = \frac{30{,}000}{62.4 \times 1.2 \times 3600} = 0.111 \text{ ft}^2$$

where S is equal to the number of tubes (n) multiplied by the flow area per tube, that is,

$$S = n \times (\pi/4)(d^2)$$

Therefore, n = 36.

Excel Solution

Water density = 62.4	d = 0.75
U = 250	Velocity = 1.2
m_c = 30,000	m_h = 15,000
Cp = 1	Cp = 1
$t_{cold-out}$ = 130	$t_{cold-in}$ = 100
Δt − cold = 30	t_{hot-in} = 200

Calculations

Q = 900,000	$t_{hot-out}$ = 140
S = 0.1113	Δt − hot = 60
n = 36.293	Δt_{LM} = 53.608
	A = 67.154

Estimate a relation between the variation of the number of tubes and the fluid velocity.

V	n
0.2	217
0.5	87
0.8	54
1.1	40
1.2	36
1.5	29
1.8	24

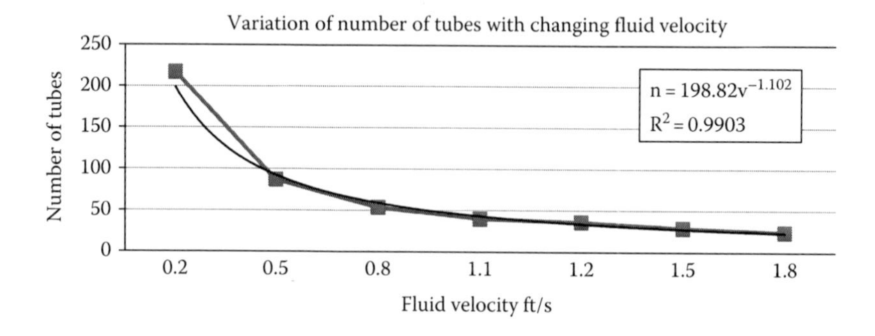

Example 6.8

An electrically heated element of surface area 12 cm² is completely immersed in a fluidized bed. The resistance of the element is measured as a function of the voltage applied to it, thus arriving at the following data:

Potential (V)	1	2	3	4	5	6
Resistance (Ω)	15.47	15.63	15.91	16.32	16.83	17.48

The relation between resistance Rw and temperature Tw is

$$\frac{Rw}{R0} = 0\ .004\ Tw - 0.092,$$

where

RO is the resistance of the wire (equal to 14 Ω) at 273 K

Tw is in K

Estimate the bed temperature and the value of the heat transfer coefficient between the surface and the bed.

Solution

The heat generation rate by electrical heating is $(V^2)/R$.

The rate of heat dissipation is hA(Tw − TB),

where Tw and TB are the wire and bed temperatures, respectively.

At equilibrium,

$$(V^2)/Rw = hA(Tw - TB)$$

But

$$\frac{Rw}{RO} = 0.004Tw - 0.092$$

so that

$$Tw = 250\left(\frac{Rw}{RO}\right) + 23$$

Thus,

$$(V^2) = \left(250hA\bar{R}w\frac{Rw}{RO}\right) - (hA\bar{R}w(TB - 23)),$$

where $\bar{R}w$ is a mean value of Rw noting that the mean cannot be used inside the bracket in the equation for Tw.

Thus, a plot of (V^2) against Rw should yield a line of slope = $250hA\bar{R}w/RO$ as given in the next diagram, from which the value of the slope is found to be 17.4.

Hence,

$$h = \frac{17.4 \times 14}{250 \times 12 \times 10^{-4} \times 16.5} = 49.2\ \text{W/m}^2\text{K}$$

The bed temperature is found by the intercept at $(V^2) = 0$, that is, when

$$Rw = 15.4\ \Omega$$

Thus,

$$TB = 250(15.4/14) + 23 = 298\ \text{K}$$

$$\text{Area} = 12\ \text{cm}^2$$

$$\text{RO at 273 K} = 14\ \Omega$$

Potential (V)	Resistance (Rw)	(V^2)
1	15.47	1
2	15.63	4
3	15.91	9
4	16.32	16
5	16.83	25
6	17.48	36

At equilibrium, $(V^2)/Rw = hA(Tw - TB)$

$$\frac{Rw}{R0} = 0.004Tw - 0.092$$

so that

$$Tw = 250\frac{Rw}{R0} + 23$$

Thus,

$$(V^2) = \left(250hA\bar{R}w\frac{Rw}{R0}\right) - (hA\bar{R}w(TB - 23))$$

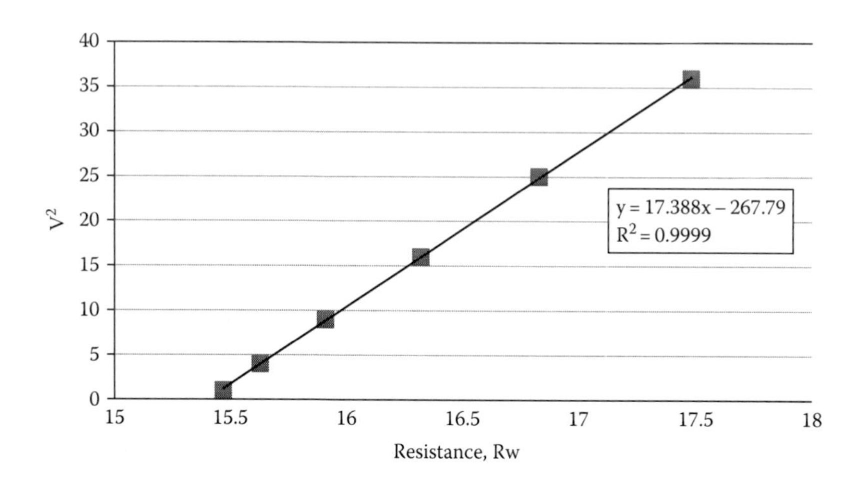

where $\bar{R}w$ is a mean value of Rw noting that the mean cannot be used inside the bracket in the equation for Tw.

Thus, a plot of (V^2) against Rw should yield a line of slope = $250hA\bar{R}w/R0$. This is shown in the figure from which the value of the slope is 17.4

$$h = 49.21\ W/m^2K$$

The bed temperature is found by the intercept at $(V^2) = 0$ that is when Rw = 15.4 Ω

$$Rw = 15.4\ \Omega$$

$$T_B = 298\ K$$

Example 6.9

Write a program to estimate the physical properties for water in the range of temperatures from 273 to 323 K.

The density for water in the range of

$$\rho = 1200.92 - 1.0056T + 0.001084T^2$$

The conductivity $K = 0.34 + 9.278 * 10^{-4}T$
The specific heat $CP = 0.015539(T - 308.2)^2 + 4180.9$

Note: Take 101 points of temperatures

Solution

```
T = 273:.5:323;
%The density
P = 120,092 – 10,056* T + 0001084* T²;
%The conductivity
K = 034 + 9278* 10⁻⁴* T;
%The specific heat
CP = 0015539* (T – 3082)² + 4180.9;
A = [T',P',K',CP']
```

6A APPENDIX: INTERNAL-FLOW CONVECTION CORRELATIONS

This workbook computes the Nusselt number for forced convection in a circular pipe as a function of the Reynolds (based on diameter) and Prandtl numbers (and where appropriate one or two other parameters). It includes subroutines for laminar, transition, and turbulent flows and for liquid metals. Results for a range of Reynolds and Prandtl numbers are shown in this plot. This spreadsheet was developed to aid in verifying our internal-flow module (Ribando, 1998).

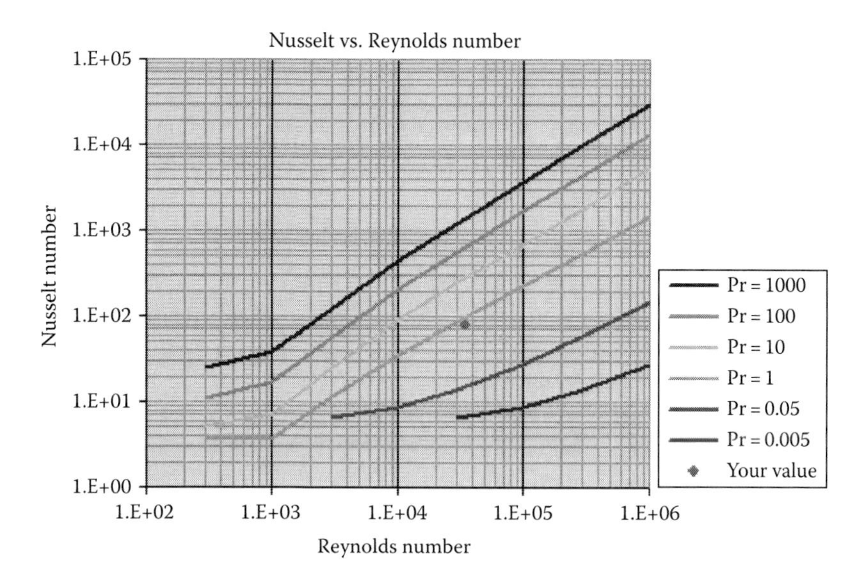

REFERENCE

Ribando, R.J. An excel/visual basic for applications (VBA) primer, *Computers in Education Journal*, VIII(2): 38–43, April–June 1998 [A version of this article updated for Excel 2007].

7 Two-Phase Gas–Liquid Diffusional Operations

Distillation and Absorption

Liquid–gas contacting systems, in general, are utilized for transferring mass, heat, and momentum between the phases, subject to constraints of physical and chemical equilibrium.

Emphasis is placed in this chapter on the *diffusional transfer of mass* between phases, known as *interphase operations.* The transfer of mass from one phase to another is involved in typical liquid–gas *mass transfer operations* described as follows:

- *Distillation (fractional distillation*) is the most widely used separation process in the chemical industry. Distillation (also known as fractional distillation) is the separation of a mixture into its component parts, or fractions. Separating chemical compounds by their boiling point by heating them to a temperature at which one or more fractions of the compound will vaporize is distillation, where the constituents or the component of the mixture boil at less than 25°C from each other under a pressure of 1 atm.
- *Flashing* is a single-stage distillation process in which the total vapor removed is kept in intimate contact with the liquid, allowing for equilibrium to be established between the vapor and liquid.
- *Rectification* is the separation of the constituents of a liquid mixture by successive distillation (implying vaporization and condensation). Separation into effectively pure components may be obtained.
- *Absorption* is the transfer of a solute component in a gas-phase stream into a liquid absorbent (solvent).
- *Stripping* (*desorption*) is the transfer of gas, dissolved in a liquid, into a gas stream.

Distillation and absorption, as two main unit operations, are covered in this chapter.

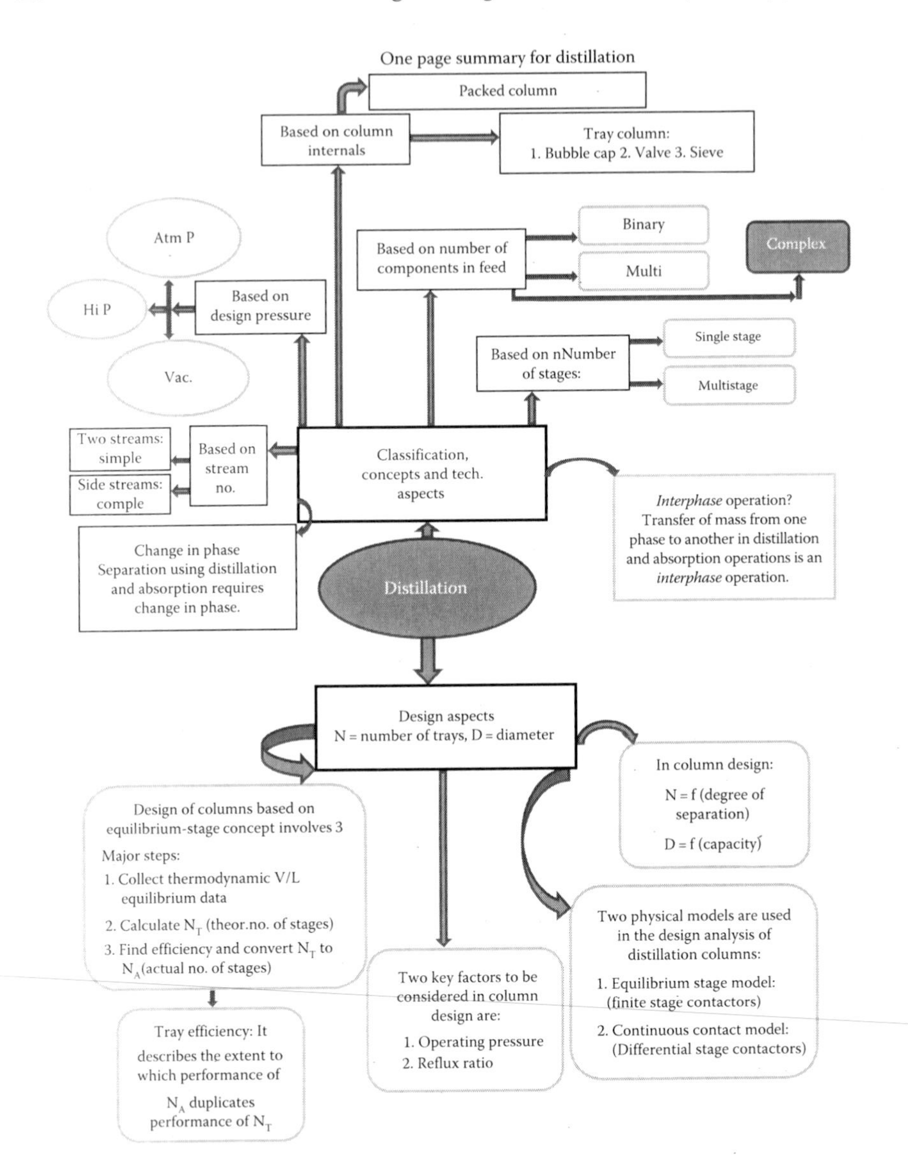

7.1 PART I: DISTILLATION

7.1.1 Introduction

7.1.1.1 Basic Concepts and Principles

7.1.1.1.1 General

- If the mixture to be separated is a homogeneous, single-phase solution, a second phase must generally be formed before separation takes place. This is an *interphase* operation, which involves the transfer of mass from one phase to another. This second phase is introduced by two methods:

- By adding or removing energy using an energy-separating agent, for example, in distillation
- By introducing a solvent using mass-separating agent, for example, in absorption

- *Intraphase* separation, on the other hand, implies separation of components within a phase, such as diffusion through inert barriers or membranes. These are rate-governed operations.
- Separation of components from a liquid mixture via distillation depends on the differences in boiling points of the individual components. Also, depending on the concentrations of the components present, the liquid mixture will have different boiling point characteristics. Therefore, distillation processes depend on the *vapor pressure* characteristics of liquid mixtures.
- For separation to take place, say by distillation, the selection of an exploitable chemical or physical property difference is very important. Factors influencing this are
  - The physical property itself
  - The magnitude of the property difference
  - The amount of material to be distilled
  - The relative properties of different species or components—purity required
  - The chemical behavior of the material during distillation and its corrosiveness.
- A measure of the ease of separation of one component i from another j is expressed by
  - The relative volatility, defined as

$$\alpha = \frac{(y_i/x_i)}{(y_j/x_j)} = \frac{K_i}{K_j}$$

where

α is the relative volatility of the more volatile component i to the less volatile j

y_i is the vapor–liquid equilibrium concentration of component i in the vapor phase

x_i is the vapor–liquid equilibrium concentration of component i in the liquid phase

y_j is the vapor–liquid equilibrium concentration of component j in the vapor phase

x_j is the vapor–liquid equilibrium concentration of component j in the liquid phase

$y_i/x_i = K$ is commonly called the *K value* or *vapor–liquid distribution ratio* of a component i

Thus, if the relative volatility between two components is very close to one, it is an indication that it will be difficult to separate the two components via distillation.

- The separation factor, $SF = [(C_i/C_j)_{\text{Top Product}}]/[(C_i/C_j)_{\text{Bottom Product}}]$ where C is the concentration of a component. High value of SF means an easy separation. A good example is the separation of salt from seawater by *evaporation*. Here, the value of SF is found, by intuition, to be infinity, because we are separating water (volatile component) from salt (nonvolatile).
- Net mass transfer between phases can occur only as long as driving force, such as a concentration difference, exists between the phases.
- The principal function of equipment used for mass transfer operations is to allow for efficient contact between the phases. Finite-stage contactors (plate columns) and continuous contactors (packed columns) are typically used.
- Different types of distillation methods are as shown next:

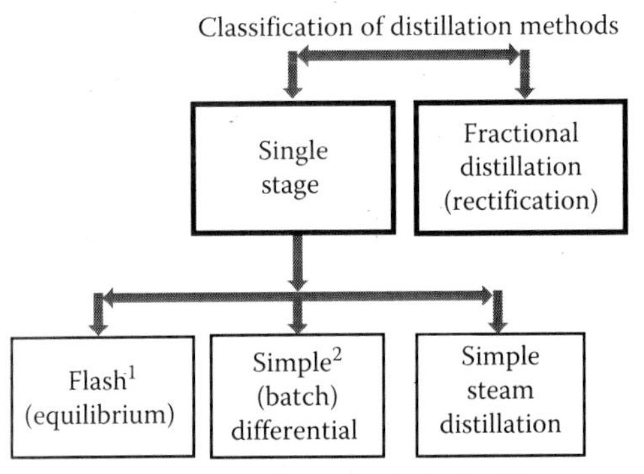

[1] Vapors are kept in intimate contact with the liquid.
[2] Vapors are withdrawn as quickly as they are formed to be condensed by a condenser.

- A stage in distillation column may be defined as a unit of equipment in which two dissimilar phases are brought into intimate contact with each other and then are separated. The contact allows for the *diffusing* components of the feed mixture to redistribute themselves between the phases. It is said that the resultant two phases have approached equilibrium.
- In an *equilibrium stage*, on the other hand, the two phases are well mixed to allow for the establishment of thermodynamic equilibrium between the phases leaving the stage, that is, V_n is in equilibrium with L_n for stage n as shown in the next diagram (notice that the arrows drawn for both streams belong to tray n). At equilibrium stage, no further change in composition of the phases occurs at a given operating conditions. To prove this concept, take the streams V_n and L_n and let them mix again for some time; examine the composition of the streams leaving. If no change is observed, then *tray* is named an equilibrium stage, or *theoretical* stage.

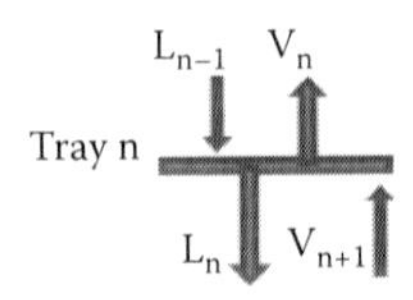

7.1.1.1.2 Design Aspects

- There are two main factors that govern the design of equipment in diffusional operations:
  - The thermodynamic equilibrium distribution of the components between the phases
  - The rate of movement, diffusion rate, from one phase to the other
- A study of the overall assembly for *finite-stage contactors* is best made on the basis of the flow and mass transfer characteristics in each individual stage. Thus, for plate columns, each plate is considered a separate entity. The overall design requires an analysis of the stepwise operation in the column from tray to tray. In a *differential-stage contactors* (packed columns) on the other hand, the contacting operation is considered as occurring continuously throughout the column. It is a continuous contact operation.
- The main factors to be considered in the design of finite-stage columns, other than calculating the number of the theoretical stages (plates) required for a given separation are the following:
  - Column diameter
  - Tray efficiency
  - Pressure drop across the tray
- The number of plates in a column is a function of the degree of separation required, that is,

$$N = f(\text{separation})$$

On the other hand, the diameter of a column is a function of the charge input to the column or capacity, that is,

$$D = f\ (\text{capacity})$$

7.1.2 Three Pillars for Distillation

Distillation models are based on three pillars:

- Laws of conservation of mass and energy
- The concept of ideal stage
- Raoult's law and Henry's law used (for ideal case) to describe the tendency of escape for vapor/liquid at equilibrium.

The law of conservation of energy (based on first law of thermodynamics) is the basis for the following fundamental heat balance for a distillation column. For 1 lb of fluid flowing through a distillation system,

$$\Delta H + \Delta K.E. + \Delta P.E. = Q - W_s$$

Usually in distillation, ΔK.E. and ΔP.E. are negligible. Therefore,

$$\Delta H = Q$$

This means that the change in enthalpy ($H_{out} - H_{in}$) of 1 lb of mass flowing is equal to the heat exchanged by the system, per 1 lb of mass flowing.

A heat balance around the distillation column gives

$$\text{Heat}_{in} = \text{Heat}_{out}$$

$$Fh_F + Q_R = DH_D + BH_B + Q_C$$

Rearranging the equation gives

$$Q_R - Q_C = [DH_D + BH_B] - FH_F$$

$$Q_{net} = \text{enthalpy of products} - \text{enthalpy of feed}$$

or

$$Q_{net} = \Delta H$$

where

ΔH is the enthalpy change
ΔK.E. is the change in kinetic energy
ΔP.E. is the potential energy change
Q is the heat added to the system
W_s is the shaft work done by the system.

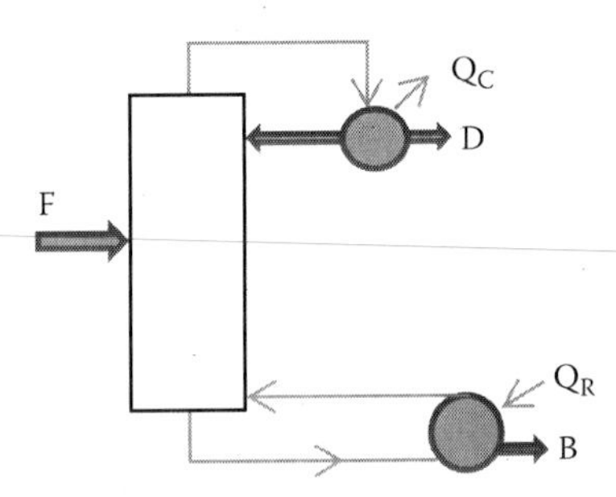

7.1.3 Features of Distillation Units

In this section, we introduce to the readers some descriptive features of the distillation units. An industrial distillation column consists of a set of trays (plates). It is classified into two sections, known as the rectification and stripping sections. The product leaving the top of the column is called the *overhead product* (distillate). Distillate product may be liquid or vapor (or occasionally both) depending on the type of condenser used, while the product leaving the bottom of the column is called the *bottom product*. Figure 7.1 illustrates some of these features.

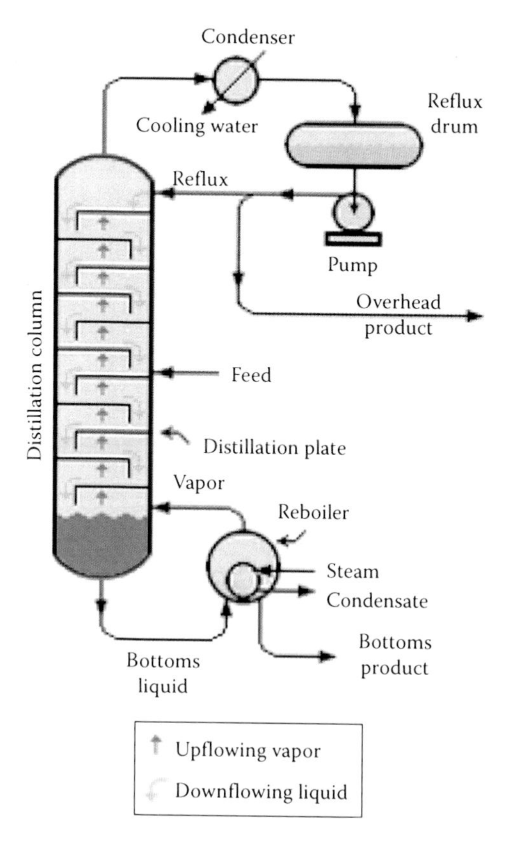

FIGURE 7.1 Schematic diagram of a distillation column.

7.1.3.1 Types of Distillation Columns

Distillation columns can be classified according to *five* criteria: function to be carried out, number of components in the feed, number of streams of products, design pressure in the column, and column internals as shown in Figure 7.2.

7.1.3.2 Overhead Condensers

There are two main categories of condenser and they differ in function by the extent of condensation:

1. A total condenser, where the entire vapor leaving the top of the column is condensed. Consequently, the composition of the vapor leaving the top tray y_1 is the same as that of the liquid distillate product and reflux, x_D.
2. A partial condenser, where the vapor is only partially liquefied. The liquid produced is returned to the column as liquid, and a vapor product stream is removed. The compositions of these three streams

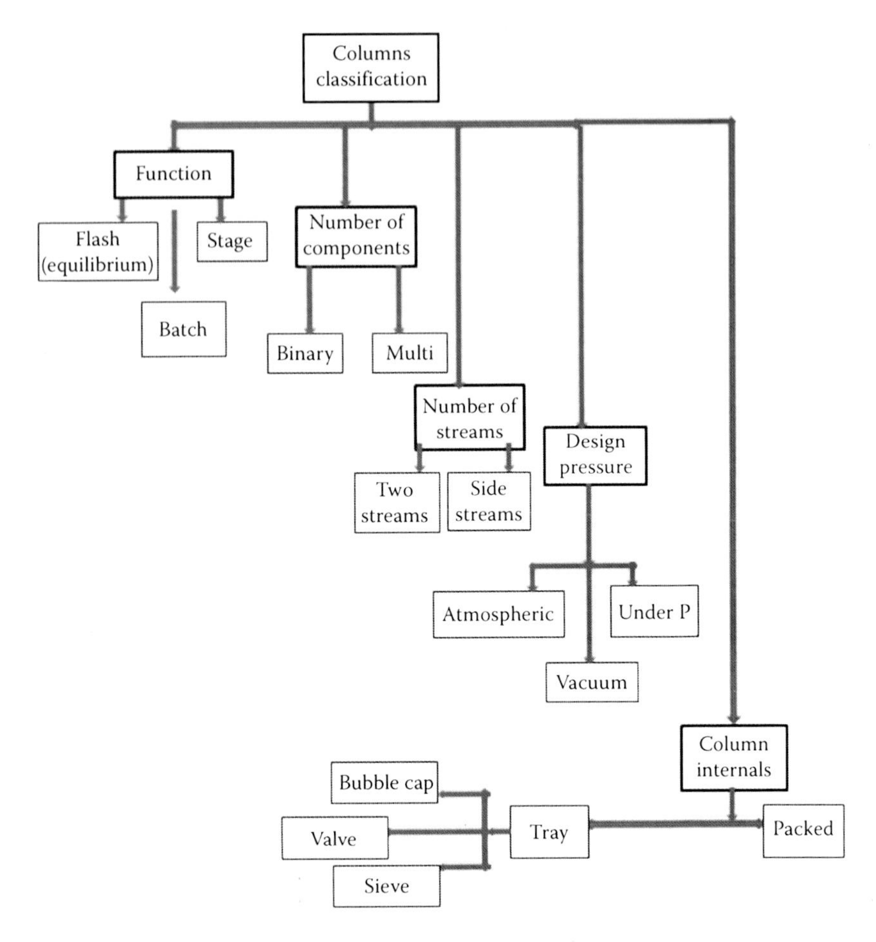

FIGURE 7.2 Classification of distillation by types of columns.

(V_1, D, and R) are different. Normally, D (composition y_D) is in equilibrium with R (composition x_D).

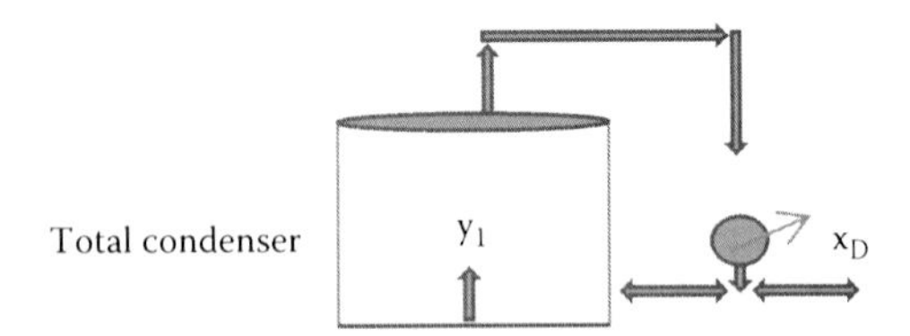

A partial condenser functions as an equilibrium separation stage, so columns with a partial condenser effectively have an extra ideal stage.

7.1.3.3 Reflux and Reflux Ratio: Top of Column

Reflux refers to the portion of the overhead liquid product returned to the upper part of the tower. The liquid reflux provides cooling and partial condensation of

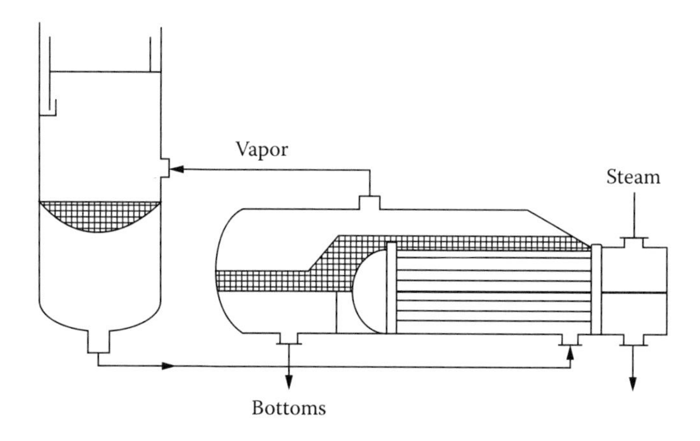

FIGURE 7.3 Schematic of kettle-type reboiler.

the upflowing vapors, thereby increasing the efficacy of the distillation tower. The *reflux ratio* is an important parameter in column operation. It is normally defined as the ratio of reflux to distillate (L/D).

7.1.3.4 Reboilers: Bottom of Column

A reboiler can be regarded as heat exchangers that are required to transfer enough energy to bring the liquid at the bottom of the column to boiling point, as shown in Figure 7.3. Most reboilers are *partial reboilers*, since they only vaporize part of the liquid in the column base. Partial reboilers also provide an ideal separation stage.

The balance of heating with a reboiler at the bottom of a column and cooling by condensed reflux at the top of the column maintains a temperature gradient along the height of the column.

7.1.3.5 Rectification and Stripping Sections in Distillation Columns

In the rectification section, vapors are enriched by removing heavy components by reflux from top, while in the stripping section, liquid stream is stripped out of light components by the uprising stripping stream or by the reboiler vapor output.

7.1.3.6 Effect of Operating Pressure

One of the primary considerations for operating pressure is the cooling medium available for the reflux condenser. The overhead product will be at bubble point conditions for a *liquid product* or at dew point conditions for *a vapor product*. These conditions are fixed by two factors:

- The desired component separation
- The temperature of the cooling medium

The cooling media typically used are *air, water, and refrigerant.*

Air cooling is the least expensive. With cooling water, process temperatures of 95°F–125°F are possible. Below 95°F, mechanical refrigeration must be used to achieve the desired condensing temperature. It is the most expensive method.

TABLE 7.1
Types and Features of Distillation Operations

Operation Features	Atmospheric Distillation	Vacuum Distillation	Pressure Distillation
Application	Fractionation of crude oils	Fractionation of heavy residues (fuel oil)	Fractionation and/or separation of light hydrocarbons
Justification	Always work near atmospheric pressure	To avoid thermal decomposition	To allow condensation of the overhead stream using cooling water
Extra equipment (as compared with atmospheric distillation)		Steam jet ejectors and condensers to produce and maintain vacuum	Stronger thickness for the vessel shell
Extra design features (as compared with atmospheric distillation)		Larger diameter because of higher vapor flow rate	Increased number of trays (N) because separation becomes more difficult; increased reflux ratio

Generally, it is desirable to operate at the lowest pressure possible to maximize relative volatility between the key components of the separation. However, if reducing the pressure requires a more expensive method, then it is not a desirable choice. Let us review next the types of operating distillation columns with respect to the operating (design) pressure, as illustrated in Table 7.1.

Solved Example 7.1

Propylene is to be separated from 1-butene by distillation into a distillate vapor; 90 mol% propylene. Calculate the column operating pressure assuming that the exit temperature from the partial condenser is 100°F, the minimum attainable with cooling water. Calculate the composition of the liquid reflux.

Solution

The operating pressure corresponds to dew point conditions for the vapor distillation conditions. Therefore, the composition of the reflux corresponds to the liquid in equilibrium with the vapor distillate at its dew point. At the dew point,

$$\sum x_i = \frac{\sum y_i}{K_i} = 1.0$$

Thus, the DP is a function of the K_i values, that is, DP = $f(\sum y_i/K_i)$ and $[(\sum y_i/K_i) - 1]$ goes to zero at the DP.

Since K_i is a function of the operating pressure, P, we can say that

$$f(P) = \left(\frac{\sum y_i}{K_i}\right) - 1$$

The method of false position (FP) will be used to perform the iterative calculations.
Now, we can say that

$$P^{k+2} = P^{k-1} - f(P^{k+1}) \times \left\{ \frac{(P^{k+1} - P^{(k)})}{[f(P^{k+1}) - f(P^{k})]} \right\}$$

To initialize calculations by the FP, two values are assumed for P: 100 and 190 psia.
As an example, calculations at 100 psia are done:

$$f(P) = \left[\left(\frac{\Sigma y_i}{K_i} \right) - 1 \right] = \left[\left(\frac{0.9}{1.97} \right) + \left(\frac{0.1}{0.675} \right) \right] - 1.0 = -0.395$$

where, 1.97 and 0.675 are the k values for propylene and 1-butene at 100 psia, respectively.

Similar iteration for next trial at P = 190 psia would lead to better values of P. The very final result will give an operating pressure of about 186 psia for the condenser outlet. At this pressure, the k values of propylene and 1-butene are 1.178 and 0.424, respectively.

The liquid composition consists of 0.764 propylene and 0.236 1-butene.

7.1.4 Design of Distillation Columns

7.1.4.1 Introduction

Fractional distillation towers or columns are designed to achieve the required separation efficiently. Two key factors should be considered before any design calculations can be made on a fractionation problem:

1. *Operating pressure*: One of the primary considerations for the operating pressure is the cooling medium available for the reflux condenser. The overhead product will be at the bubble point conditions for *a liquid product* or at dew point conditions for *a vapor product*. In return, the bubble point or the dew point pressure is fixed by a desired component separation and the temperature of the cooling medium.

   The cooling media typically used are air, water, and refrigerant. Cooling by air is considered the least expensive. Practical exchanger design limits the process to a 20°F approach to the ambient summer temperature. This means that a process temperature of 115°F–125°F would apply. With cooling water, process temperature of 95°F–105°F is applicable. However, for process temperature below 95°F, refrigeration must be applied for cooling.
2. *Reflux ratio and minimum number of stages* (*trays*): The design of a fractionating column could be looked upon as a *capital cost* versus an *energy cost* trade-off issue, which is basically a function of

   - The number of stages
   - The reflux ratio

A distillation column can only accomplish a desired separation between the two limits of minimum reflux (MR) and minimum number of trays, as depicted in Figure. 7.4 Neither of these situations represents real operations.

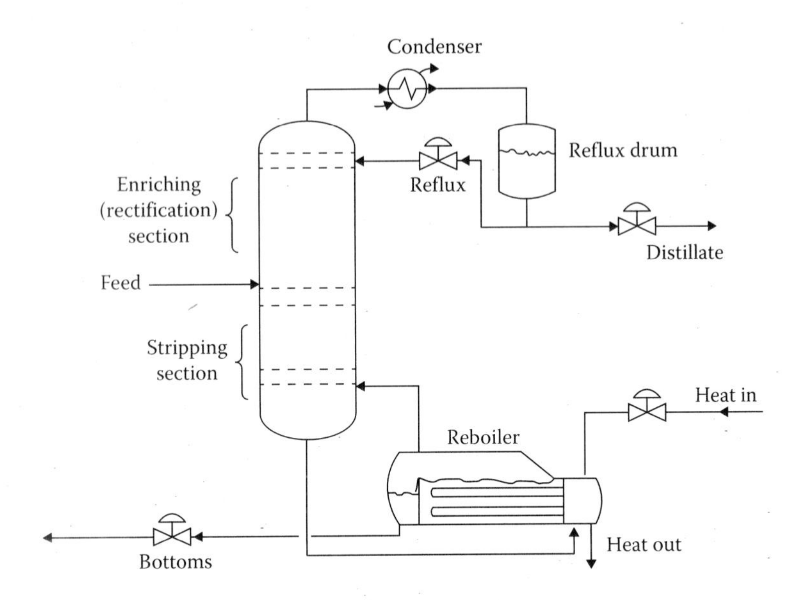

FIGURE 7.4 Rectification and stripping sections in distillation.

Consideration of these two key factors is what we may call the process design stage of distillation column. On the other hand, the purpose of the mechanical design is to select the tower internals and column diameter and height. Some of the factors involved in design calculations include feed load size and properties and the type of distillation column utilized.

7.1.4.2 Approach to Solve Distillation Problems

There are two types of problems in distillation:

1. *The design problem*
   - *Given*: The desired separation (purity)
   - *Find*: The number of trays required in a column in order to separate a given feed into products of desired purities

   For a binary distillation, we would usually specify the mole fraction of the more volatile component in the distillate and bottom products. The column pressure, feed flow rate, thermal condition (e.g., temperature, fraction vapor), and composition are also specified. In addition, the (external) reflux ratio is usually specified.
2. *The rating (operational) problem*
   - *Given*: The number of trays in a column
   - *Find*: The performance (separation)

   There are a number of different types of rating problems. All of them involve iterative (trial-and-error) solution. A guess is made of one or more unknown variables, and a design solution procedure is carried out using the values guessed. The results are compared with the column specifications, and if they do not match, a new trial is made and the solution is repeated.

7.1.4.3 Physical Models for Distillation

Two physical models are used in the design analysis of distillation columns:

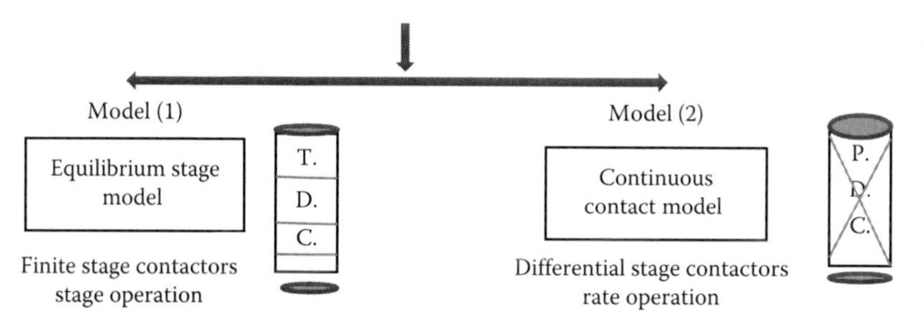

Model (1)	Model (2)
1. To establish a standard for measuring the performance, the ideal stage concept is introduced. The approach to equilibrium realized in any stage is defined as stage efficiency η	1. In this model, equilibrium is never allowed to be achieved between phases at any point. Should equilibrium occur anywhere in the system, this will be equivalent to an ∞ number of trays.
2. The design of tray column would involve three basic steps: (1) VLE data (2) Calculation of N_{theor} (3) Finding the value of η to calculateN_{act}	2. In the design of packed column, the efficiency η, is function of the transport rates of the species and the construction of the separating device. N_{theor} is calculated by a differential balance on the column

7.1.4.4 Calculation of Number of Trays

The initial calculations involved in column design are based on relatively simple stoichiometric (material balance) and equilibrium relationships.

The design of a distillation column depends upon three parameters:

- The composition of the feed
- The thermal condition of the feed
- The composition of the desired products

Methods applied to calculate the number of trays for distillation columns are usually grouped into the following categories:

Feed	Recommended Solution
• Binary systems	Graphical methods, for example, McCabe–Thiele
• Multicomponent mixtures	Shortcut methods
• Complex mixtures (petroleum hydrocarbons)	Rules of thumb (empirical)

7.1.5 McCabe–Thiele Method

7.1.5.1 Introduction

The method is basically a graphical design technique for stagewise contactors. Its utility stems from the fact that the design involves the simultaneous solution of equilibrium relationships (VLE) and the operating line to determine the number of theoretical stages (trays) required to achieve a desired degree of separation. It is a simplified method of analysis making use of several assumptions, yet it is a very useful tool for the understanding of distillation operation. The VLE data must be available at the operating pressure of the column.

The information required is the feed condition (temperature, composition), distillate and bottom compositions, and the reflux ratio. Detailed procedure and solved example are given next.

7.1.5.2 Assumptions

- Constant molar overflow.
- Molar heat of vaporization of the components is roughly the same.

7.1.5.3 Procedure

1. The equation of the operating line for the rectification section is derived by making a material balance on two passing streams on a tray, above the feed.

$$y = \left(\frac{L}{V}\right)x + \left(\frac{D}{V}\right)x_D$$

$$= \left[\frac{R}{(R+1)}\right]x + \left[\frac{1}{(R+1)}\right]x_D$$

where

y is the mole fraction of the more volatile component in the vapor phase
x is the mole fraction of the more volatile component in the liquid phase
x_D is the mole fraction of the more volatile component in the distillate
D is the distillate product
L is the flow rate of liquid stream in the rectifying section
V is the flow rate of vapor stream in the rectifying section
R is the reflux ratio L/D

This is an equation of straight line, which is drawn as follows. The desired top product composition is located on the VLE diagram on the x-axis, and a vertical line is drawn until it intersects the diagonal line. A line with slope = R/(R + 1) is then drawn from this intersection point as shown in the diagram later (Figure 7.5).

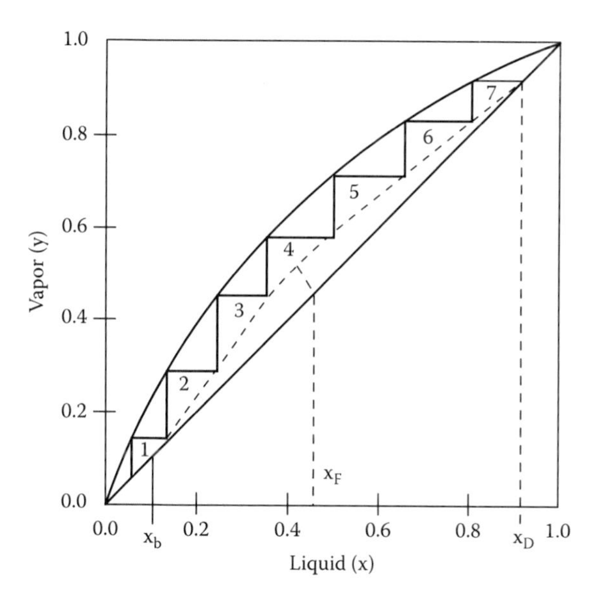

FIGURE 7.5 McCabe–Thiele diagrams for calculating N.

2. The equation of the operating line for the stripping section is derived using a similar approach for a tray below the feed.

$$y = \left(\frac{L}{V}\right)x - \left(\frac{B}{V}\right)x_B$$

   where

   y is the mole fraction of the more volatile component in the vapor phase
   x is the mole fraction of the more volatile component in the liquid phase
   x_B is the mole fraction of the more volatile component in the bottom product
   B is the bottom product
   L is the flow rate of liquid stream in the rectifying section
   V is the flow rate of vapor stream in the rectifying section

   The operating line for the *stripping section* is then drawn in a similar manner. Here, the starting point is the desired bottom product composition. A vertical line is drawn from this point to the diagonal line, and a line of slope L/V is drawn.

3. Given the operating lines for both the stripping and the rectification sections, the graphical construction of the steps is applied. In this way, we count the number of steps as shown in the diagram. This particular example shows that the seven *theoretical* stages are required to achieve the desired

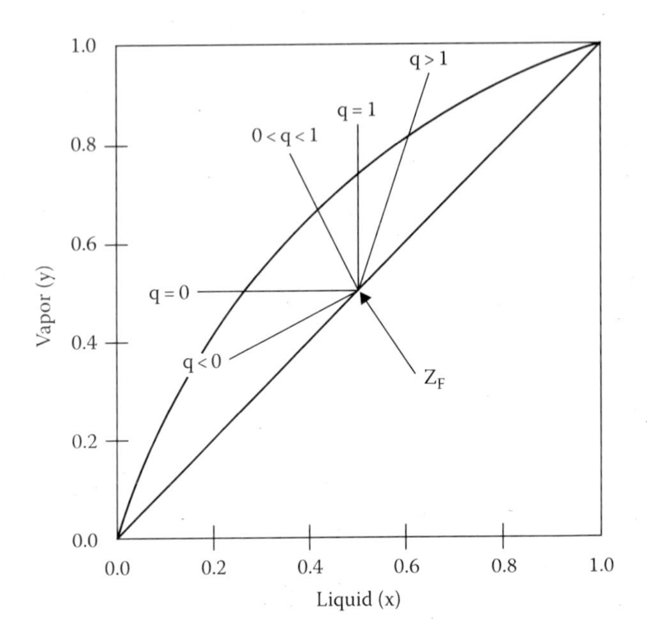

FIGURE 7.6 Feed q-lines.

separation. The required number of trays (as opposed to stages) is one less than the number of stages since the graphical construction includes the contribution of the reboiler in carrying out the separation.

The actual number of trays required is

$$N_{actual} = \frac{N_{theorit}}{\eta}$$

where η stands for efficiency.

4. Location of the feed line (q-line) is determined by the conditions (state) of the feed as shown next (Figure 7.6). In this example, feed is introduced on the fourth tray.

```
q = 0 (saturated vapor)
q = 1 (saturated liquid)
0 < q < 1 (mix of liquid and vapor)
q > 1 (subcooled liquid)
q < 0 (superheated vapor)
```

7.1.5.4 Illustration of How to Use Excel to Solve a Problem

This Excel spreadsheet uses the McCabe–Thiele method to calculate the number of theoretical stages needed for binary distillation. Binary distillation is a common unit of operation that separates two liquids (with one being more volatile or *lighter*). The concepts are encountered in many branches of chemical and petroleum engineering.

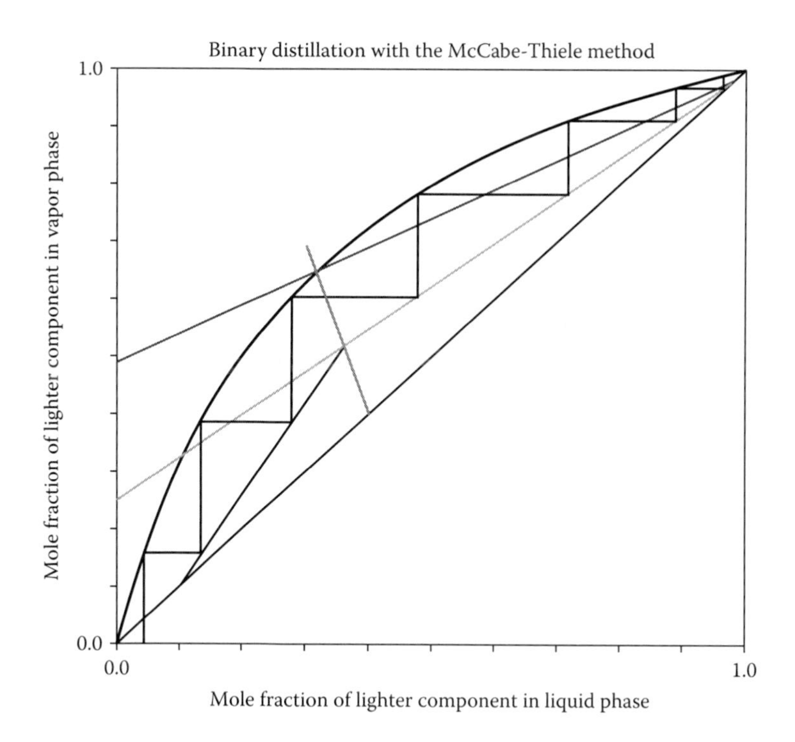

Simply enter your parameters at the top of the spreadsheet, including the

- Feed flow rate
- Mole fraction of the light component (or more volatile component) in the feed
- Mole fraction of the light component in the top product
- Mole fraction of the light component in the bottom product
- Reflux ratio
- Relative volatility of the light component
- q-line value

The spreadsheet will automatically calculate flow rates throughout the column, the number of theoretical plates, the feed plate position, the MR ratio (from the Underwood equation), and the minimum number of theoretical plates (from the Fenske equation). (Microsoft Office Excel, 2007).

Example 7.2: Plate-to-Plate Calculations (*Case of Rectification Column*)

Concept

Thiel and Geddes method is used to calculate the number of trays. It involves the simultaneous solution of equilibrium relationships (VLE) and the operating line, where the operating line is used to compute the composition of one of the two streams passing each other for two consecutive plates, while the equilibrium relationship is used to compute the composition of either the vapor or liquid (in equilibrium) on the same plate.

This example is an oversimplified one for illustrative purposes.

Given

1. Derivation of the operating line

$$y_{n+1} = \left[\frac{R}{R+1}\right]X_n + \left[\frac{1}{R+1}\right]X_D. \quad (7.2.1)$$

where R is the reflux ratio (RR) = L/V and X_D is the composition of the overhead product.

2. Equilibrium data

$$X_n = \frac{Y_n}{[Y_n + \alpha(1 - Y_n)]} \quad (7.2.2)$$

Statement of the Example

Given: 40 mole/h of feed (*vapor*) that contains 20% hexane and 80% octane entering the bottom plate, where D = 5 mole/h, $X_D = 0.9$, R.R = 7, and $\alpha = 6$. Find the number of theoretical trays, *N.1*st: first:first

First: Numerical Solution

Steps:

1. The liquid composition leaving the partial condenser (plate number 0) is in equilibrium with the vapor (top product) and is calculated by Equation 7.2.2. Hence, X_{reflux} (leaving plate 0) = 0.9/.0.9 + 6(1 − 0.9) = 0.6
2. Y_1 is calculated using Equation 7.2.1; substituting for R = 7, X = 0.6, and $X_D = 0.9$, we get $Y_1 = 0.637$
3. Get the equilibrium composition of the liquid on the same tray, $X_1 = 0.226$
4. Again, using Equation 7.2.1, get $y_2 = 0.3$
5. Next, get $X_2 = 0.066$, which is the bottom product leaving the column; call it X_w

Finally, we make overall MB, and CMB:

$$(40)(0.2) = (5)(0.9) + (35)X_w \Rightarrow$$

Solve for $X_w = 0.1$

Therefore, two plates plus the condenser make a total of *three theoretical plates*.

Second: Using Excel

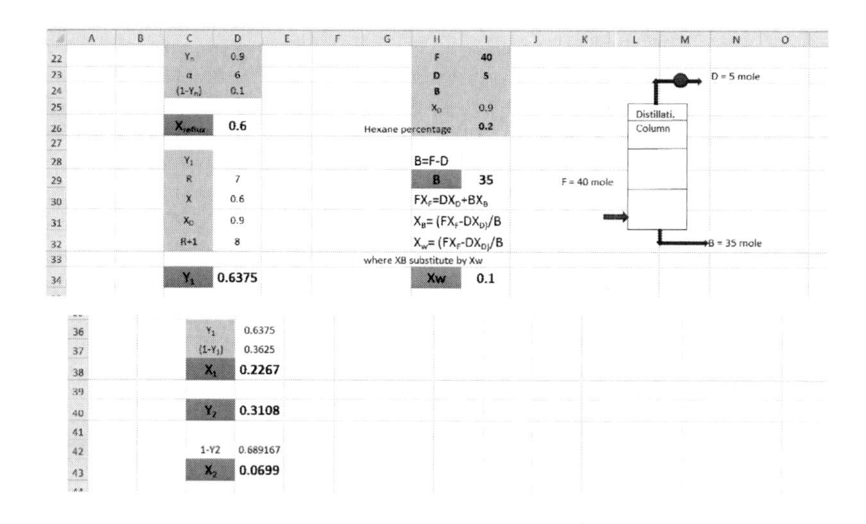

Example 7.3: Plate-to-Plate Calculations by Excel (*Case of Stripping Column*)

A liquid mixture at the boiling point that consists of 70 mole% benzene and 30 mole% toluene is fed to a stripping column. Pressure is taken 1 atm. Feed rate is 400 kg mole/h. Stripping operation is carried out to achieve a bottom product W = 60 kg·mol/h that contains no more than 2 mole% benzene.

Solve the problem using Excel, in order to determine the number of theoretical trays, N, required to obtain the desired specifications of the bottom product W. Use α_{AB} relative volatility for benzene/toluene, where $\alpha_{AB} = K_{Benzene} / K_{Toluene} = P_B^0 / P_T^0 = 2.45$

Solution

The number of trays required to reach a bottom product, exiting the stripping column, is found to be around 11 trays, as seen next; that corresponds to 1.9 mole% benzene.

					BENZENE mol%	TOLUENE mol%	
F	FEED	Lm	400	Kg mol/hr	70.0	30.0	100.0
W	BOTTOM		60	Kg mol/hr	10.0	90.0	100.0
D	TOP	Vm	340	Kg mol/hr T.M.B	80.6	19.4	100.0

Trayes	vapor Y	liquied X
1	0.806	0.629
2	0.757	0.560
3	0.677	0.461
4	0.560	0.342
5	0.420	0.228
6	0.286	0.140
7	0.183	0.084
8	0.116	0.051
9	0.078	0.033
10	0.057	0.024
11	0.046	0.019
12	0.040	0.017

Operating line constants	
Lm/Vm	1.1765
Constant	0.0176

Relative volatility	2.45

Design of distillation column
Plate-to-plate calculation case of striping column- binary system
By
GHADA EZZ ELDIN ALLAM
Under supervision of
DR. HUSSEIN ABDEL-AAL

7.1.5.5 Total and Minimum Reflux: Limiting Cases

In distillation analysis, separation of a pair of components can be improved by increasing the number of stages while holding reflux constant or by increasing the reflux flow for a given number of stages. This trade-off sets up two limiting cases:

- *Total reflux* (TR) with minimum ideal stages, N_m
- *MR* with infinite ideal stages, N_∞

The design trade-off between reflux and stages is the standard economic optimization problem chemical engineers always face: balancing the capital costs for the number of trays *versus* the operating cost for the reflux to be recirculated. A good design will operate near a cost optimum reflux ratio, as illustrated in Figure 7.7.

The TR condition represents operation with no product removal. The overhead vapor is condensed totally and returned as reflux. Consequently, the reflux ratio (L/D) is infinite. The operating lines coincide with the diagonal line.

The aforementioned conditions of TR and MR can be further visualized by the following approach (with the aid of the opposite diagrams):

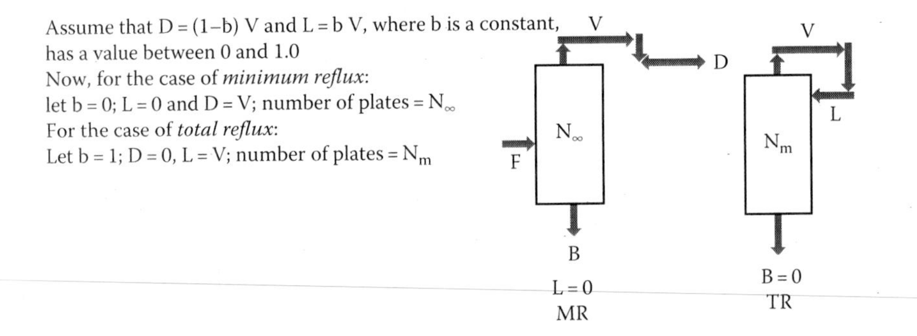

The MR condition, on the other hand, represents the theoretical opposite of TR—an infinite number of ideal separation stages. In this case, the intersection of the operating lines lies on the equilibrium curve itself.

This intersection is called a *pinch point*. A simple column will have two pinch points (because there are two operating lines). The points change when the operating lines do. A pinch at the intersection of the feed line and the equilibrium curve indicates that the column is operating at *MR*.

7.1.5.6 Tray Efficiency

Mass transfer limitations prevent the vapor leaving a tray from being in true equilibrium with the liquid on the tray. The assumption we make of ideal stages

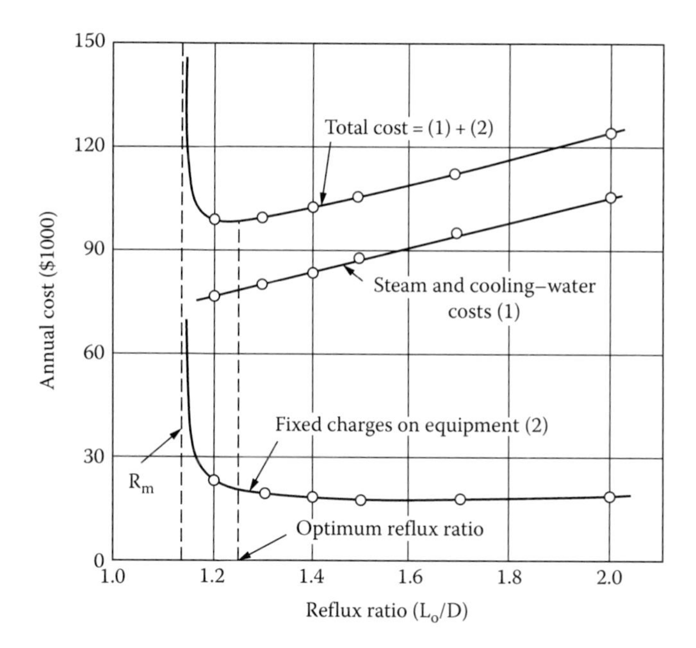

FIGURE 7.7 Optimum reflux ratio for distillation column.

is only an approximation. The concept of efficiency is introduced to represent the deviation from equilibrium. Two types of efficiencies are considered in our discussion:

- Murphree efficiency is probably the most common to use. It assumes perfect mixing on *the tray*. Values between 0.6 and 0.75 are common for sieve trays.
- Overall efficiency is the simplest choice. It is the ratio of the number of ideal stages to the number of actual stages.

One can replace the true equilibrium curve with an *effective equilibrium curve (EEC)* located between the true curve and the operating lines as shown in Figure 7.8. In this case, the number of stages counted will be the true or actual trays.

Murphree efficiency is as indicated in the following equation:

$$\text{EEC} = \frac{[Y_n - Y_{n+1}]}{[Y_n^* - Y_{n+1}]}$$

$$= \frac{Z_1}{Z_2}$$

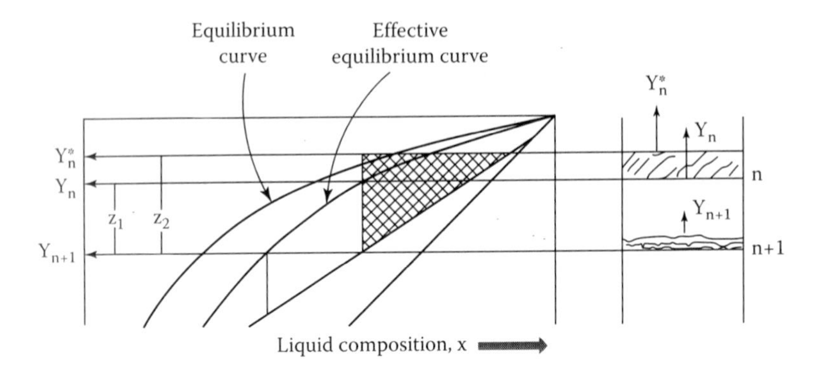

FIGURE 7.8 Illustration of an effective equilibrium curve.

O'Connell has correlated efficiency data for distillation columns as a function of

- *Liquid viscosity* (μ)
- *Relative volatility of key components* (α) (or gas solubility)

The results of the correlation for fractionators and absorbers are graphically presented by Peters and Timmerhaus. Approximate estimation of efficiency is done by a number of methods as described by Van Winkle.

7.1.5.7 Stage Design and Efficiency

Following the *equilibrium stage model* described earlier, distillation columns are designed to give discontinuous contact of phases in a series of stages.

A single-stage contact unit is designed to accomplish two main tasks:

1. To allow the contact of the V and L in such a manner, where the interfacial area and time of contact are at their maximum for the stage to approach equilibrium
2. To handle the required quantities of V and L without excessive pressure drop

The approach to equilibrium, on the other hand, is a *rate factor*, intimately tied up into the rate of mass and heat transfer. A great number of possible variables contribute to these rate processes, which makes the evaluation of efficiency extremely difficult.

Therefore, once N_T (N_{theor}) is calculated, estimation of the efficiency is carried out to find N_A (N_{act}) rather than taking the rate processes to evaluate the efficiency as indicated in Figure 7.9a.

In other words, to determine N_A, one can pursue one of the two options:

1. The one described in Figure 7.9a
2. *Head-on route*, illustrated by the next diagram (Figure 7.9b)

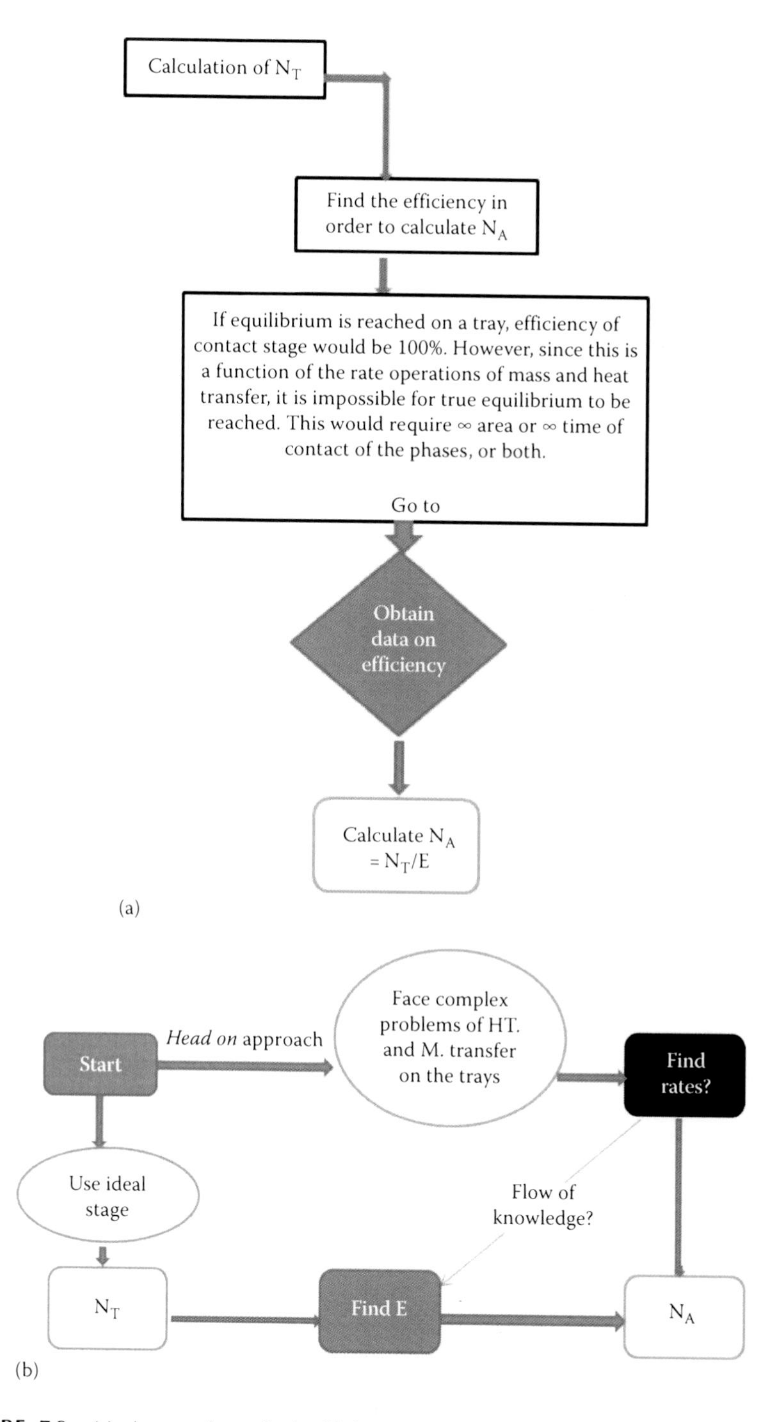

FIGURE 7.9 (a) Approach to find efficiency. (b) *Head on* approach to find the efficiency (N_T, N_A, and E are theoretical number of trays, actual number, and tray efficiency, respectively).

7.1.6 Shortcut Methods: Fenske–Underwood–Gilliland

7.1.6.1 Introduction

There are many so-called *shortcut* calculation methods for designing industrial distillation columns. The most commonly used one is the Fenske–Underwood–Gilliland method.

- The Fenske equation estimates the minimum number of theoretical plates or equilibrium stages at total reflux.
- The Underwood equation estimates the MR for an infinite number of theoretical equilibrium stages.
- The Gilliland method then uses Fenske's minimum plates and Underwood's MR to estimate the theoretical plates for a given distillation at a chosen reflux.

Shortcut calculations provided by the Fenske–Underwood–Gilliland are most effective for a preliminary design before the use of distillation simulation software that utilizes much more rigorous calculation methods.

7.1.6.2 Fenske Equation

$$N = \frac{\log\left[(X_d/(1-X_d))((1-X_b)/X_b)\right]}{\log \alpha_{avg}}$$

where

N is the minimum number of theoretical plates required at TR (of which the reboiler is one)

X_d is the mole fraction of the more volatile component in the overhead distillate

X_b is the mole fraction of the more volatile component in the bottom products

α_{avg} is the average relative volatility of the more volatile component to the less volatile component

7.1.6.3 Underwood Equation

It is represented by two relationships:

1. First equation

$$\frac{(\alpha_1)(z_{F1})}{[(\alpha_1)-\theta]} + \frac{(\alpha_2)(z_{F2})}{[(\alpha_2)-\theta]} + \cdots = 1-q$$

where

z_F is the mole fraction of component n, in the feed

q is the thermal condition of feed (heat required to vaporize 1 mol of feed divided by molar latent heat of feed)

2. Second equation

$$R_m + 1 = \frac{(\alpha_1)(x_{D1})}{[(\alpha_1) - \theta]} + \frac{(\alpha_2)(x_{D2})}{[(\alpha_2) - \theta]} + \cdots$$

where X_D is the mole fraction of component n, in the distillate.

The first equation is solved for the value of the θ, which lies between the relative volatilities of the key components.

Having obtained θ, we calculate R_m from the second equation.

7.1.6.4 Gilliland's Method

Gilliland used an empirical correlation to calculate the final number of stage N, using the diagram (Figure 7.10) as follows:

From the values calculated through the Fenske and Underwood equations (N_{min}, R, R_{min}), where R is set to $1.5R_m$, one enters the diagram with the abscissa value $[R - R_m]/[R + 1]$, which is known, and reads the ordinate of the corresponding point on the Gilliland curve. The only unknown of the ordinate is the number of *theoretical* stages.

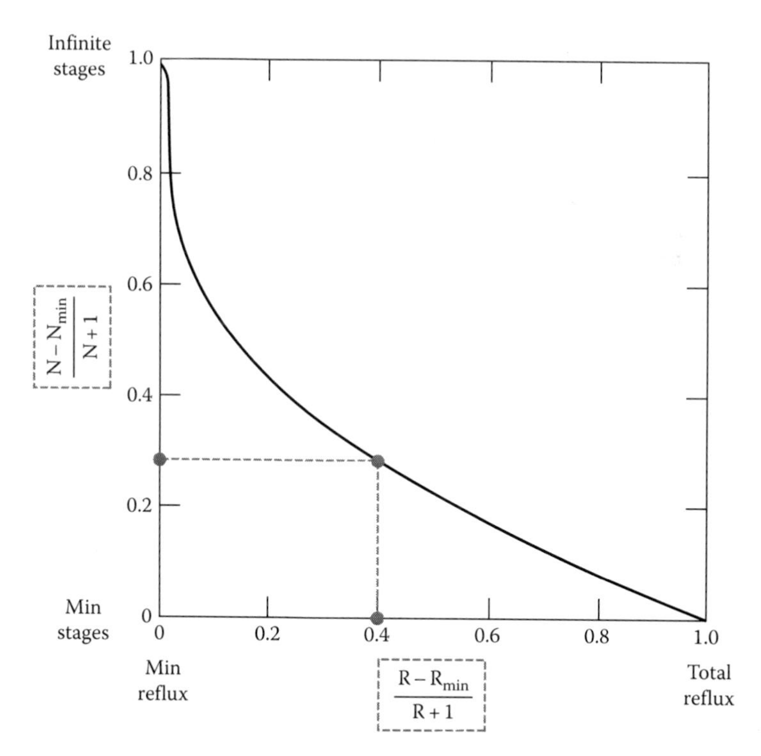

FIGURE 7.10 Gilliland correlation.

7.1.7 Column Diameter

A fractionating tower must have sufficient cross-sectional area to handle the rising vapors without excessive carry-over of liquid from tray to tray. The maximum allowable vapor velocity, V_m applicable for any specific location in the column is given by Souders and Brown as follows:

$$V_m = K_v \sqrt{\frac{[\ell_L - \ell_V]}{\ell_L}}$$

An alternative form of this equation, in terms of mass velocity is

$$G_m = V_m \rho_v = K_v \sqrt{\frac{[\rho_L - \rho_v]}{\rho_L}}$$

where

- V_m is the maximum allowable vapor velocity, based on cross-sectional area of empty tower, ft/s
- K_v is the an empirical constant, ft/s. It is a major function of tray spacing
- ρ_L and ρ_v are the liquid and vapor densities, lb/ft^3, respectively
- G_m is the maximum allowable mass velocity of vapor, lb/(s) · (ft^2)

The total vapor load in the column

$$Q\ (\text{lb/s}) = V_m\ (\text{ft/s})\, \ell_V\ (\text{lb/ft}^3)\, A\ (\text{ft}^2)$$

Solving the equation for the cross-sectional area A, the tower diameter is readily obtained.

7.1.8 Flash Distillation

The problem of separating the gas from crude oil for well fluids (crude oil mixtures) breaks down to the well-known problem of flashing a partially vaporized feed mixture into two streams: vapor and liquid. In the first case, we use a gas–oil separator; in the second case, we use what we call a flashing column.

Under the assumption of equilibrium conditions, and knowing the composition of the fluid stream coming into the separator and the working pressure and temperature conditions, we could apply our current knowledge of vapor/liquid/equilibrium (flash calculations) and calculate the vapor and liquid fractions at each stage.

A *flash* is a single-stage distillation in which a feed is partially vaporized to give a vapor that is richer in the more volatile components. This is the case of a feed heated under pressure and flashed adiabatically across a valve to a lower pressure, the vapor being separated from the liquid residue in a flash drum. This is the case of *light liquids*. Apart from the gas–oil separation problem addressed here, methods used in practice to produce and hence separate two-phase mixtures are as follows:

Initial Phase	Process and Conditions
Higher-pressure liquids (light)	Heat under pressure, then flash adiabatically using valve
Low-pressure liquids (heavy)	Partial vaporization by heating, flash isothermally (no valve)
Gas	Cool-after initial compression
Gas	Expand through a valve or engine

For flashing to take place, the feed has to be two-phase mixture, that is, it satisfies the following:

$T_{BP} < T_f < T_{DP}$ (as indicated in the VLE diagram, Figure 7.11).

Or,

The sum of (Z_iK_i) for all components is greater than 1 and the sum of [z_i/K_i] for all components is less than 1, where T_{BP}, T_f, and T_{DP} are the bubble point of the feed mixture, flash temperature, and dew point of the feed mixture, respectively.

Z_i and K_i are the feed composition and equilibrium constant, respectively, for component i.

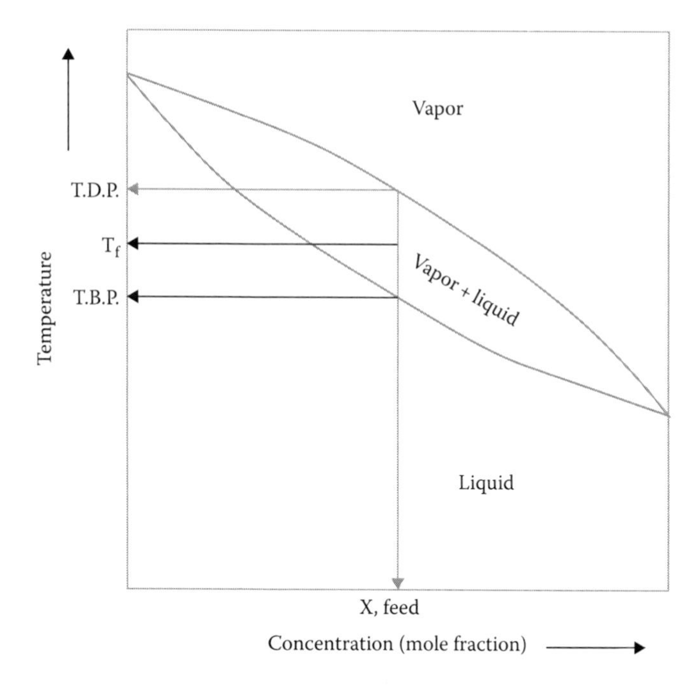

FIGURE 7.11 Conditions for the flashing of a binary system.

7.1.8.1 Flash Equation

The need to discuss flash calculation arises from the fact that it provides a tool to determine the relative amounts of the separation products V(gas) and L(oil) and their composition Y_i and X_i, respectively.

The flash equation is derived by component material balance $Fz_i = Vy_i + Lx_i$, as presented in Figure 7.12.

Two forms for the flash equation are presented:

1. *Simple form*

$$X_i = \frac{Z_i}{1-(V/F)(1-K_i)}$$

For given conditions of P and T, the solution of the equation to find the value of X_i is obtained by trial and error assuming a value for V/F (take F unity), until the sum of $[X_i] = 1$ is satisfied.

2. *Functional form*

$$f(g) = \sum_{i=1}^{i=c}\left[\frac{Z_i}{1-g(1-K_i)}\right]-1$$

where g = V/F. The desired root to this function, gr = (V/F)r, is the value that makes the function f(g) go to zero or the sum of $[X_i] = 1$. The solution of this flash equation is carried out using computers. Details on using functional technique are fully described in many references.

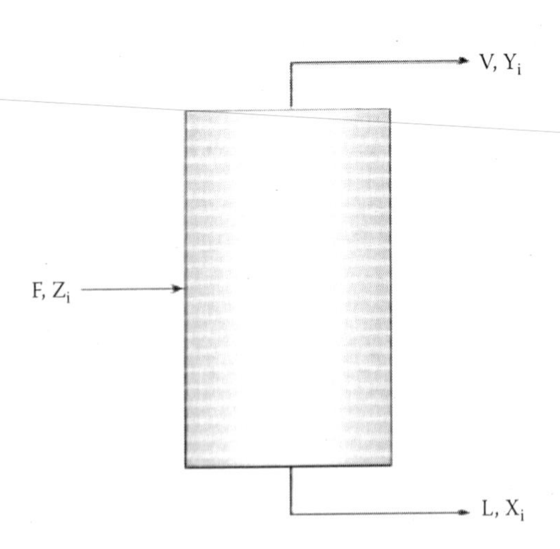

FIGURE 7.12 Flash separation of binary mixture.

The vapor–liquid equilibrium constant, defined as

$$K_i = \frac{Y_i}{X_i}$$

where

- Y_i is the mole fraction of component i in the vapor phase
- X_i is the mole fraction of component i in the liquid phase
- K_i is the equilibrium constant of component i for a given T and P, is considered the key concept used in the computation of phase behavior of hydrocarbon mixtures in oil and gas streams

K is called the *distribution constant* because it predicts the distribution of a component in each phase: vapor and liquid. K is a function of T, P, and the composition of a given system. The K values most widely used are those developed by the National Gas Processors Association. Because the K values are designated for a specific convergence pressure (defined as the pressure at a given T, where the values of K for all component in a system become or tend to become equal to unity), it is important that the value of the operating pressure of the system should be below the convergence pressure used in predicting the K values.

7.1.9 Reactive Distillation

Reactive distillation is a process where the chemical reactor is also the still. Separation of the product from the reaction mixture does not need a separate distillation step, which saves energy (for heating) and materials. A case study on reactive distillation (on the production of methyl acetate) is presented in Chapter 10. It is a manifestation of the merits of reactive distillation.

7.2 PART II: ABSORPTION AND STRIPPING

Absorption is the transfer of a solute component in a gas-phase stream into a liquid absorbent (solvent). The objective in absorption is to bring a *gas stream* into direct contact with a *liquid solvent* to enable soluble components of the gas to be washed out by the solvent. Absorption finds applications in two main fields:

- The recovery of valuable components from a gas stream
- The removal of obnoxious materials or components from a gas stream

Stripping (desorption) on the other hand, is the transfer of gas, dissolved in a liquid, into a gas stream. It is the reverse of absorption. The fluid to be processed is *liquid*;

the aim is to liberate a dissolved gas from the fluid. Stripping can be carried out by more than one option:

- Using a stream of inert gas
- Applying a temperature rise by heating and boiling of solutions
- Applying pressure reduction

Design methods for absorbers are presented. This includes graphical technique, numerical (analytical) approach, and the unit concept method.

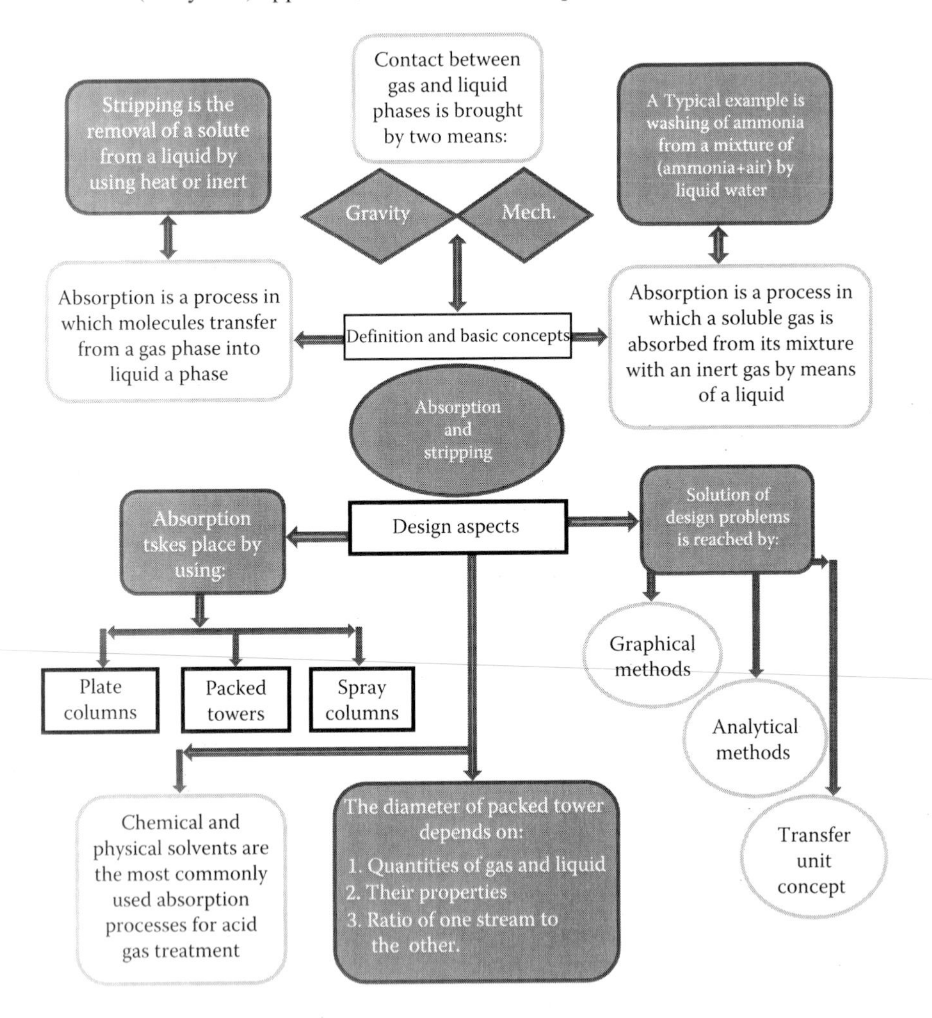

7.2.1 Basic Concepts

Absorption depends on intimate contact of gas and liquid phases under favorable conditions. Contact is achieved by maximizing the interfacial area of surfaces of gas and/or liquid to each other. This is achieved by two approaches:

The action of gravity (buoyancy) of the fluid by using:

(a) Plate towers

Breaking the gas into little bubbles passing through a volume of liquid.

(b) Packed columns

Breaking the liquid into multiplicity of slow-flowing films, which forms and reforms through a volume of gas.

(c) Wetted-wall columns

Mechanical means or devices by using:

(a) Spray columns

Breaking the liquid into tiny drops dispersed through a volume of gas.

(b) Centrifugal absorbers

The general features of absorption and stripping operations are illustrated as follows:

Absorption: Removal of c_{2+} from natural gas using n-Octane (lean O)

Stripping: Removal of light hydrocarbons from heavy cuts by heating or using inert.

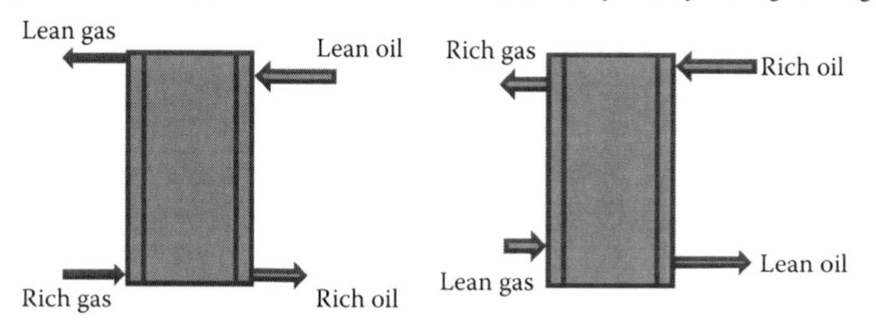

7.2.2 Outline of Design Approach

Solution of design problems in absorption is reached by one of the following methods:

1. Graphical methods
2. Analytical methods using absorption factors
3. Transfer unit concept

The following parameters are the most important to consider in the design of absorbers/strippers:

- The best solvent to use
- The optimum gas velocity through the absorber, hence the vessel diameter
- The height of the vessel and its internal members or the number of contacting trays
- The optimum solvent circulation rate through the absorber and stripper
- Temperatures of streams entering and leaving the absorber and stripper, and the quantity of heat removed
- Operating pressures at which absorption process takes place

7.2.3 Graphical Methods: Solution for Simple Countercurrent Absorption Operations

The most generally satisfactory methods of determining the number of ideal stages in a cascade are *graphical* ones. Similar to distillation, the graphical determination of the number of stages in absorption involves the alternate use of the operating line and the equilibrium curve as shown in Figure 7.13a. When the solute is being transferred from the liquid phase, L, to the vapor phase, V, the process is called *stripping*. This is case 7.13b.

The operating line could be plotted by either of the two options:

- By knowing all four of the compositions at both ends of the column
- By knowing three compositions and the slope L/V of the operating line

In absorption problems, there are *two carriers* and a *solute* which is partitioned between them. An example is the use of sulfuric acid to absorb humidity (water) from moist air. The two carriers are: air and acid, where water is the solute. This is represented by

Other absorption problems involve the use of nonvolatile solvents. An example is the scrubbing of benzene/air mixture using a nonvolatile hydrocarbon oil as solvent.

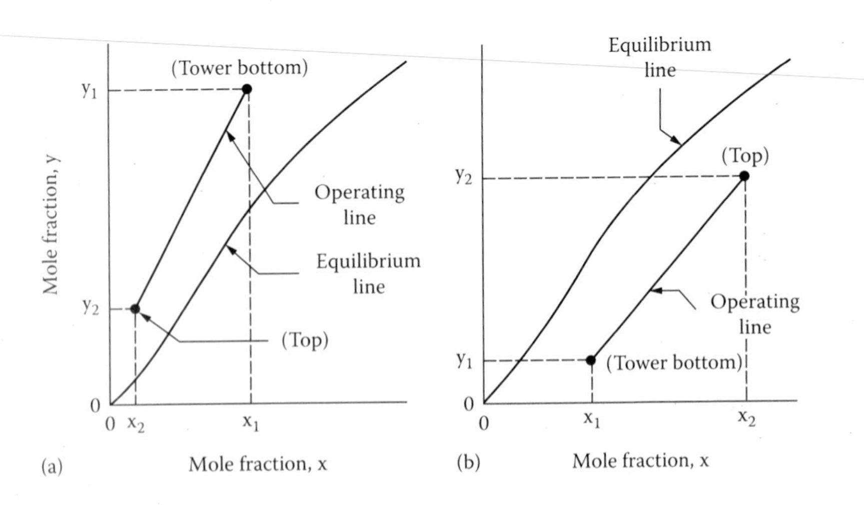

FIGURE 7.13 Illustrations for the operating line and equilibrium curve for absorbers/strippers. (a) Case of absorption. (b) Case of stripping.

7.2.4 Material Balance Calculations

In the absorber, shown in Figure 7.14, we have two streams flowing countercurrent under the following conditions:

1. The gas stream flow rate G (mol/h), at any point in the tower consists of two components:
   a. Diffusing solute, A, composition y, or $Y = y/(1 - y)$ and $G_A = G \cdot y$; and
   b. Nondiffusing (inert gas), composition is $(1 - y)$; $V_s = V(1 - y)$ and $(V_s + V_A) = [V(1 - y) + V \cdot y] = V$
2. The liquid stream, L mole/h·ft^2, at any point, is composed of two species:
   a. Nonvolatile liquid, L_s
   b. The absorbed solute of the gas, in the liquid phase, having mole fraction: $X = x/(1 - x)$ and $L_s = L(1 - x)$, $L_A = Lx$

Since V_s and L_s are almost constant in quantity, as they pass through the tower, it is convenient to express the material balance in terms of these flow rates.

Solute balance around the boundary shown in the diagram yields

$$V_s(Y_1 - Y) = L_s(X_1 - X)$$

This represents an equation of straight line (operating line):

$$\boxed{Y_1 - Y} = [L_s/V_s]\boxed{X_2 - X}\text{, or}$$

Where, at the terminal conditions: $Y \Rightarrow Y_2$ and $X \Rightarrow X_1$

where

Y is the moles of solute/moles of inert gas

X is the moles of solute/moles of inert liquid

Finally, we obtain

$$\frac{L}{V} = \frac{[y_b - y_a]}{[x_b - x_a]}$$

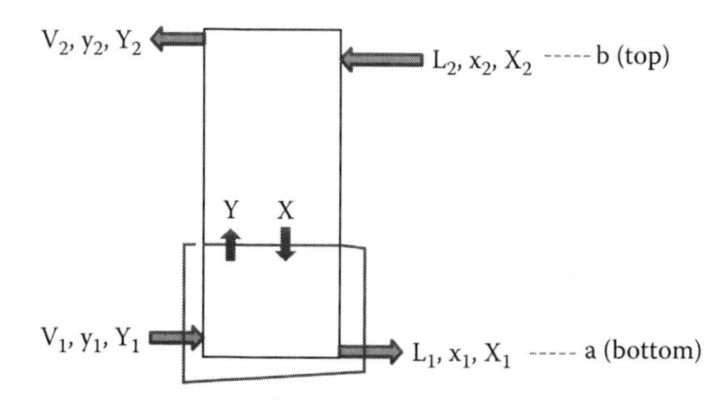

FIGURE 7.14 Material balance for an absorber.

Example 7.4

Moist air that enters is to be dried in a countercurrent drier using sulfuric acid to absorb the moisture. Conditions are as given next.

Given

(i) Moisture content in air is $Y_b = 28 \times 10^{-3}$ moles water/mol dry air.
(ii) Water content is $Y_a = 8 \times 10^{-3}$ after drying.
(iii) The acid introduced to the dryer has a composition of $X_a = 2$ moles water/mol (H_2SO_4) and leaves with $X_b = 9$. With reference to the given equilibrium diagram,

Find

(a) The ratio of liquid absorbent to gas treated
(b) The smallest ratio of acid to air that can be used to dry the air from point "b" to point "a" shown in Figure 7.15

Solution

(a) As shown earlier, the slope of the operating line EF is
Y/X = ratio of: H_2SO_4 Acid to Air treated
$= 2.86 \times 10^{-3}$, mole acid to mole dry air
$= 2.86$ mole acid for 1000 moles dry air
(b) Line FH has the lowest slope through F that can provide a value for $Y_b = 2.8 \times 10^{-2}$.

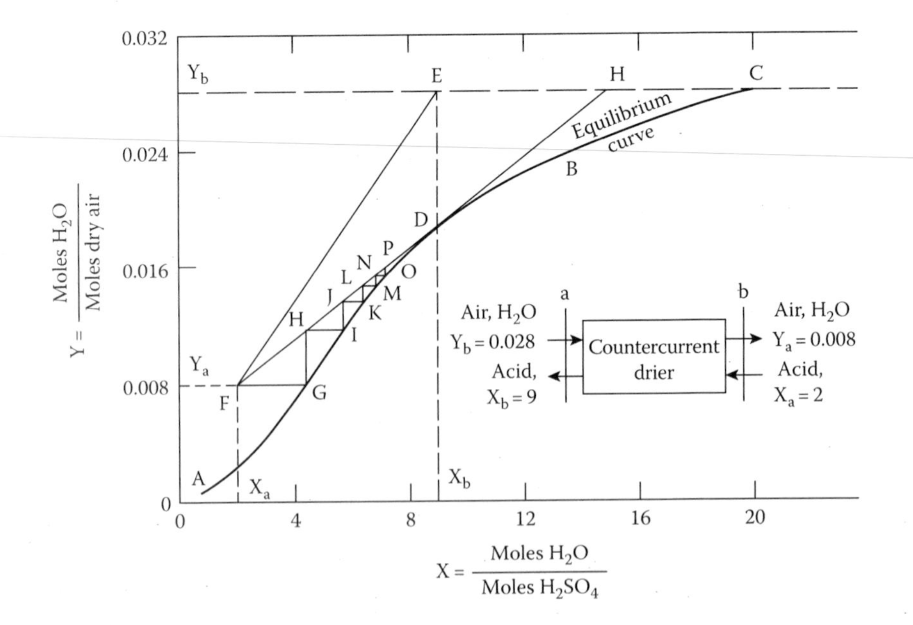

FIGURE 7.15 Graphical solution of Example 7.4.

This line is a tangent to the equilibrium curve at point D. It has a slope of 1.64×10^{-3} mol acid per mole air. Although this is the smallest amount of acid to be used, it requires an infinite number of stages as shown in the diagram (pinch point).

Example 7.5

Using nonvolatile hydrocarbon oil as a solvent, benzene/air mixture is to be scrubbed using a simple countercurrent absorption column. Data are shown in Figure 7.16.

Given

- Inlet feed gas contains 5% benzene.
- Flow rate of feed gas is 600 lb moles/h.
- Solubility of benzene in oil follows Raoult's law.
- The average molecular weight of oil is 200.
- Tower operates isothermally at 80°F and a pressure of 1 atm.

Find

(a) The minimum oil rate (lb/h) required to recover 90% of the entering benzene
(b) The number of theoretical stages needed if an oil rate that is 1.5 times the minimum is used.

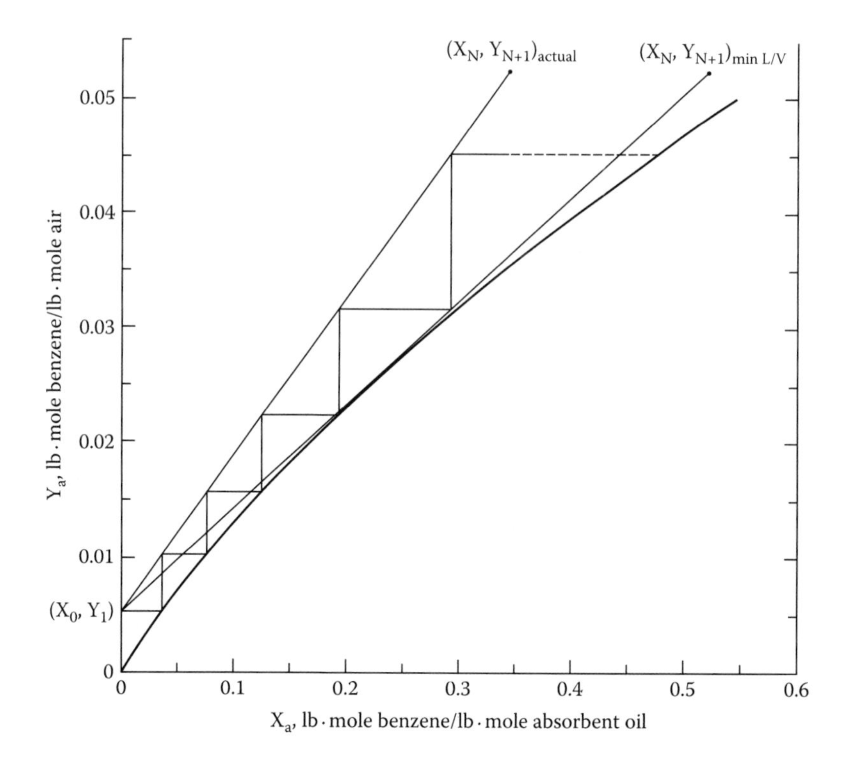

FIGURE 7.16 Graphical solution of Example 7.5.

Solution

First. Calculation of the equilibrium data, Y_a/X_a, for benzene/air is done.

For benzene,

$$Y_a = \left(\frac{P_a}{P}\right) x_a$$

At 80°F, $P_a = 103$ mm Hg and P = 760 mm Hg; therefore, $y_a = 0.136x_a$.

To assure a straight operating line, mole ratios will be used: $y_a = Y_a/(1 + Y_a)$ and $x_a = X_a/(1 + X_a)$; hence, $Y_a/(1 + Y_a) = 0.136\ X_a/(1 + X_a)$.

Assuming values for Y_a in the range from 0 to 0.05, the corresponding values of X_a are calculated and the equilibrium data of Y_a and X_a are plotted as shown in Figure.

Inlet gas contains (0.05)(600) = 30 lb mol of benzene. Recovery of benzene is 90%. Benzene leaving in the gas is 3 lb mole; $Y_1 = 3.0/570 = 0.00526$ and $X_0 = 0$.

Conditions at the upper end (outlet) are now set and plotted. Feed conditions are

$$X_0 = 0.0 \quad \text{and} \quad Y_{N+1} = \frac{0.05}{0.95} = 0.0526$$

hence, the operating line is drawn.

Minimum liquid rate occurs when the operating line touches the equilibrium curve to give

$$\left(\frac{L}{V}\right)_m = \frac{(0.0526 - 000526)}{(0.52 - 0)} = 0.091$$

$$(\text{Oil rate})_m = (0.091)(570)(200)\ \text{lb/h}$$

Operating oil rate = $(1.5)(\text{Oil rate})_m = 15{,}600$ lb/h, and the value of $X_N = 0.345$

The number of stages are figured out to be about five stages for the given oil rate.

A detailed discussion for the design methods of plate and packed absorbers is beyond the scope of this chapter. Reference to *Perry's Handbook* (50th edition), pp. 14-14–14-24 is recommended. Heat effects in gas absorption, multicomponent systems, and absorption with chemical reaction are fully discussed as well on pp. 14-24–14-31.

7.2.5 Analytical Methods: Equations to Calculate the Number of Stages for Absorbers/Strippers

As stated earlier, when the operating line and the equilibrium lines are both straight over a given concentration range, the number of stages N can be calculated by formula (analytically), as shown next. In this case, graphical technique is unnecessary. The method is known as the *absorption factor method.* Derivation of the absorption factor equation is detailed in McCabe and Smith. The final derived equations are

1. For the transfer of solute "A" from vapor phase to liquid phase, that is, *absorption*, the following relationship is obtained:

$$N = \frac{\log\{[(y_b - y_b^*)/(y_a - y_a^*)]}{\log A}$$

When A, the absorption factor, is defined as L/mV = 1,

$$N = \frac{[y_b - y_a]}{[y_a - y_a^*]}$$

where m is the slope of the operating line.

2. For the *stripping* process, when S is defined as 1/A = mL/V

$$N = \frac{[x_a - x_b]}{[x_b - x_b^*]}$$

These series of equations are known as *Kremser* equations.

Example 7.6

Ammonia is to be scrubbed from a gas containing 4% ammonia and 96% air by volume, in a countercurrent plate column. Water containing 0.003 mole NH_3 per mole of water is used at a rate of 1.1 mole H_2O per mole of air. Ninety percent of the ammonia entering is to be scrubbed. K factor for ammonia is

$$= 0.8\ [\text{mole}\, NH_3/\text{mole air}]/[\text{mole}\, NH_3/\text{mole}\, H_2O].$$

Calculate:

(a) The exit concentration of ammonia in the liquid
(b) The number of ideal trays required

Solution

(a) Material balance based on 1 mole of air and using mole ratio units is

$$Y_b = \frac{4}{96} = 0.04167$$

Ammonia absorbed = 90% [0.04167] = 0.0375 = 0.0341 mol ammonia per mol of H_2O.

$X_b = 0.003 + 0.0341 = 0.0371$ mole ammonia per mole of H_2O.

(b) $Y_a = 0.00417$ mole ammonia per mole of air; $A = 1.1/0.8 = 1.375$; $Y_a^* = 0.00334$; and $Y_b^* = 0.02968$

Therefore,

$$N = \frac{\log\{[0.04167 - 0.02968]/[0.00417 - 0.00334]\}}{\log 1.375}$$

$$= 14 \text{ trays}$$

7.2.6 Transfer Unit Concept for Absorption in Packed Columns

The following is a brief discussion for the calculation of the height of a packed absorption tower, Z. Because of space limitation, presentation is given for the case of transfer unit based on gas film:

$$Z = (H_G)(N_G)$$

where

H_G is the height of a transfer unit based on the gas film. The units of H are in meter

N_G is the number of transfer units

When the solutions in gas absorption are dilute (concentrations < 10%), the following equation is applicable:

$$Z = H_G N_G = H_G \left[\frac{(1-y)_{LM}}{(1-y)}\right]_{av} \left[\frac{(y_1 - y_2)}{(y - y_i)}\right]$$

where

y_1 and y_2 are the mole fractions of component A in gas inlet and gas outlet, respectively

y and y_i are the mole fraction in bulk of gas phase and at gas side of interface, respectively

7.2.7 Role of Absorption in Natural Gas Field Processing

Natural gas associated with crude oil production or produced from gas fields generally contains undesirable components such as H_2S, CO_2, N_2, and water vapor. Field processing of natural gas implies the removal of such undesirable components before the gas can be marketed.

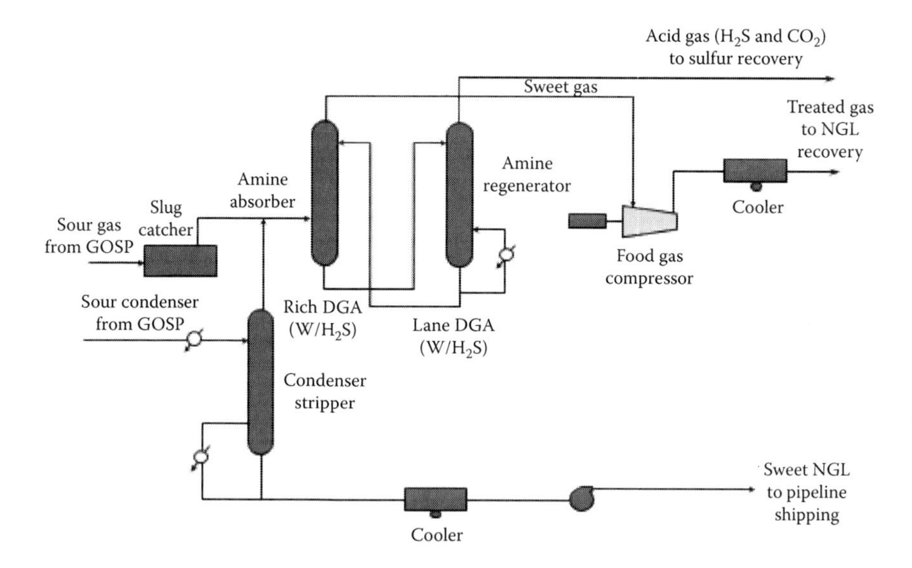

FIGURE 7.17 Separation of natural gas liquid by absorption.

Chemical and physical solvents are the most commonly used absorption processes for acid gas treatment. In the chemical process, mono- and diethanolamine are used to react with the acid gases reversibly and form products that can be generated by a change in temperature as given in Figure 7.16 The physical process, on the other hand, uses a solvent to absorb the sour gases (Sulfinol process) from the natural gas. The absorbed impurities are then separated from the rich solvent by steam stripping.

Another important application for gas absorption is the separation of the components of natural gas liquid, known as C_{2+}, ethane and heavier. This is done using heavy hydrocarbon cut to absorb these components, which are generated from the rich solvent by distillation, as shown in Figure 7.17.

END-OF-CHAPTER SOLVED EXAMPLES

Example 7.7

For the following distillation column, write a code to find the value of stream B and the compositions of stream B.

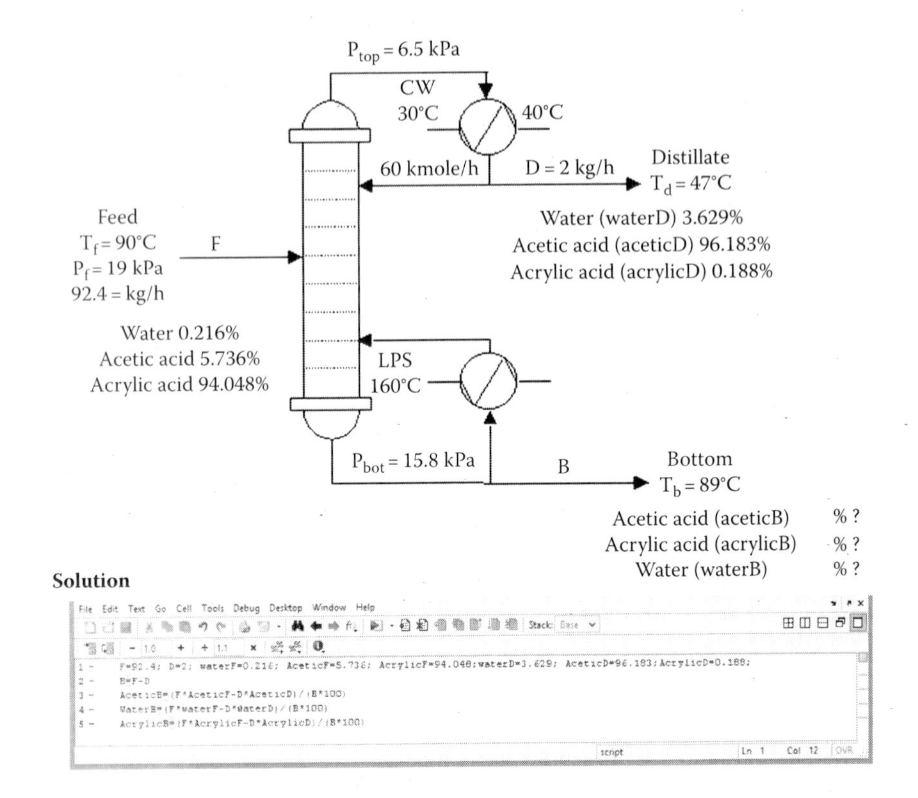

Solution

```
1 -  F=92.4; D=2; waterF=0.216; AceticF=5.736; AcrylicF=94.048;waterD=3.629; AceticD=96.183;AcrylicD=0.188;
2 -  B=F-D
3 -  AceticB=(F*AceticF-D*AceticD)/(B*100)
4 -  WaterB=(F*waterF-D*WaterD)/(B*100)
5 -  AcrylicB=(F*AcrylicF-D*AcrylicD)/(B*100)
```

MATLAB®

Results

```
Command Window
New to MATLAB? Watch this Video, see Demos, or read Getting Started.

B =

   90.400000000000006


AceticB =

   0.037349601769912


WaterB =

   0.001404911504425


AcrylicB =

   0.961245486725664
```

Example 7.8

Xylene, styrene, toluene, and benzene are to be separated with the array of distillation columns that are shown later. Write a program to calculate the amount of the streams D, B, D_1, B_1, D_2, and B_2 and also to calculate the composition of streams D and B.

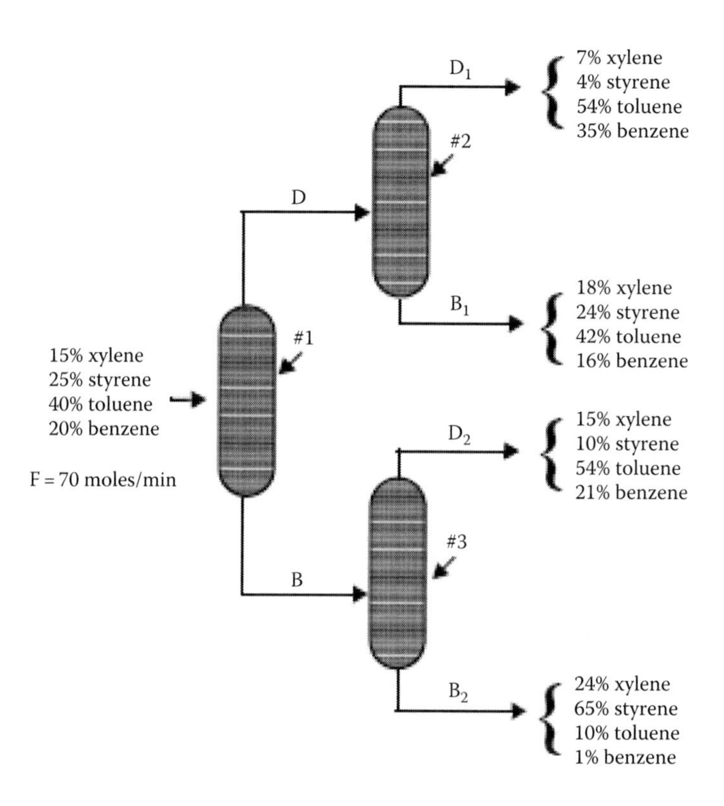

Solution

Making material balances on individual components on the overall separation train yields the following equation set:

Xylene: $0.07D_1 + 0.18B_1 + 0.15D_2 + 0.24B_2 = 0.15 \times 70$
Styrene: $0.04D_1 + 0.24B_1 + 0.10D_2 + 0.65B_2 = 0.25 \times 70$
Toluene: $0.54D_1 + 0.42B_1 + 0.54D_2 + 0.10B_2 = 0.40 \times 70$
Benzene: $0.35D_1 + 0.16B_1 + 0.21D_2 + 0.01B_2 = 0.20 \times 70$

Overall material balances and individual component balances on column 2 can be used to determine the molar flow rate and mole fractions from the equation of stream D.

Molar flow rates: $D = D_1 + B_1$
Xylene: $XDxD = 0.07D_1 + 0.18B_1$
Styrene: $XDsD = 0.04D_1 + 0.24B_1$
Toluene: $XDtD = 0.54D_1 + 0.42B_1$
Benzene: $XDbD = 0.35D_1 + 0.16B_1$

where
XDx is the mole fraction of xylene
XDs is the mole fraction of styrene
XDt is the mole fraction of toluene
XDb is the mole fraction of benzene

Similarly, overall balances and individual component balances on column 3 can be used to determine the molar flow rate and mole fractions of stream B from the equation set.

Molar flow rates: $B = D_2 + B_2$
Xylene: $XBxB = 0.15D_2 + 0.24B_2$
Styrene: $XBsB = 0.10D_2 + 0.65B_2$
Toluene: $XBtB = 0.54D_2 + 0.10B_2$
Benzene: $XBbB = 0.21D_2 + 0.01B_2$

where F, D, B, D_1, B_1, D_2, and B2 are the molar flow rates in mole/min.

Solution

```
clear all and clc

A =[0.07,0.18,0.15,0.24; 0.04,0.24,0.10,0.65;0.54,0.42,0.54,0.1;0.35,0.16,0.21,0.01];
B=[0 .15*70; 0 .25*70; 0 .4*70; 0 .2*70];
X=A\B; D1=X(1), B1=X(2), D2=X(3), B2=X(4), D=D1+B1
B=D2+B2

XDx=(.07*D1+.18*B1)/D XDs=(.04*D1+.24*B1)/D
XDt=(.54*D1+.42*B1)/D XDb=(.35*D1+.16*B1)/D
XBx=(.15*D2+.24*B2)/B XBs=(.1*D2+.65*B2)/B
XBt=(.54*D2+.1*B2)/B XBb=(.21*D2+.01*B2)/B
```

The results will be

```
D1  = 26.2500
B1  = 17.5000
D2  = 8.7500
B2  = 17.5000
D   = 43.7500
B   = 26.2500
XDx = 0.1140
XDs = 0.1200
XDt = 0.4920
XDb = 0.2740
XBx = 0.2100
XBs = 0.4667
XBt = 0.2467
XBb = 0.0767
```

Example 7.9

For the following separation system, we know the inlet mass flow rate (in kg/h) and the mass fractions of each species in the inlet (stream 1) and each outlet (streams 2, 4, and 5). Calculate the unknown mass flow rates of each outlet stream.

Solution

If we define the unknowns as `x1 = F1`, `x2 = F2`, and `x3 = F3` and set up the mass balances for the following:

1. The total mass flow rate `x1 + x2 + x3 = 10`
2. The mass balance on species 1
   `0.04x1 + 0.54x2 + 0.26x3 = 0.2*10`
3. The mass balance on species 2
   `0 .93x1 + 0 .24x2 = 0 .6*10`

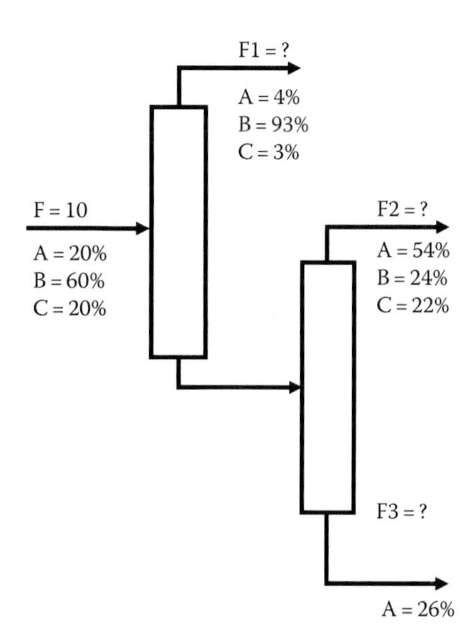

then these three equations can be written in matrix form

$$\begin{bmatrix} 1 & 1 & 1 \\ 0.04 & 0.54 & 0.26 \\ 0.93 & 0.24 & 0 \end{bmatrix} \begin{bmatrix} x1 \\ x2 \\ x3 \end{bmatrix} = \begin{bmatrix} 10 \\ 2 \\ 6 \end{bmatrix}$$

To find the values of unknown flow rates, write the code

```
A = [1,1,1;.04,.54,.26;.93,.24,0]; B = [10;.2*10;.6*10];
X = AB; F1 = X(1),F2 = X(2),F3 = X(3)
```

The results will be

```
F1 = 5.8238
F2 = 2.4330
F3 = 1.7433
```

Example 7.10

In an absorption tower, ammonia is absorbed from air at atmospheric pressure by acetic acid. The flowrate of 2 kg/m²s in a test corresponds to a Reynolds number of 5100 and hence a friction factor $R/\rho u^2$ of 0.020. At the temperature of absorption the viscosity of the gas stream is 0.018 mN·s/m², the density is 1.154 kg/m³, and the diffusion coefficient of ammonia in air is 1.96×10^{-5} m²/s.

a) Determine the mass transfer coefficient through the gas film in kg/m²s (kN/m²).
b) Determine the mass transfer coefficient through the gas film in kg/m²s (kN/m²) if the flow rate increases from 2 to 10 kg/m²s.

Solution

Given

Flow rate = 2.00 kg/m²s
Reynolds no. = 5100
Friction factor $R/\rho u^2$ = 0.020
Viscosty = 0.0180 mN·s/m²
Density = 1.154 kg/m³
Diffusion coefficient = 0.000020 m²/s

(a)

```
[(hd/u)*(PBm/P)*(u/pD)exp0.56]=jd
(u/pD)exp0.56 = 0.88
(hd/u)*(PBm/P)= 0.02261364
u = G/p = 1.73310225
KG = [(hd/R*T)*(PBm/P)] = 1.5319E-05
KG = 0.00026392 kg/m²s (kN/m²)
KG = 2.70 x 10-4 kg/m²s (kN/m²)
```

(b)

The previous steps were repeated for different ammonia flow rates and the mass transfer coefficient was calculated in each case.

The results were shown in the following table and were plotted in the figure.

Flow Rate	kg
2	0.0002
4	0.0005
6	0.0008
8	0.00107
10	0.00134

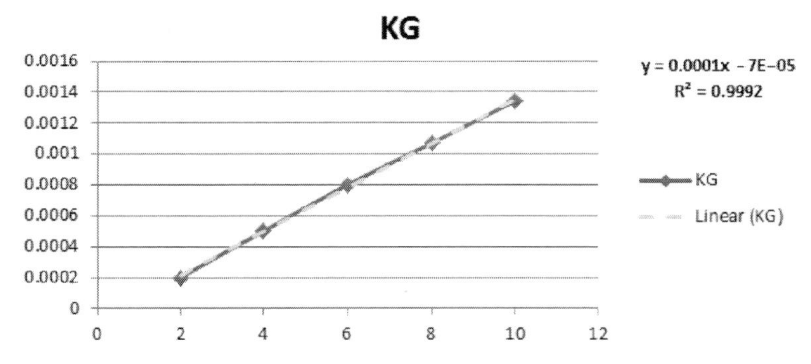

Example 7.11

A continuous fractionating column is required to separate a mixture containing 0.695 mole fraction n-heptane (C_7H_{16}) and 0.305 mole fraction n-octane (C_8H_{18}) into products of 99 mole % purity. The column is to operate at 101.3 kN/m^2 with a vapor velocity of 0.6 m/s. The feed is all liquid at its boiling-point, and this is supplied to the column at 1.25 kg/s. The boiling point at the top of the column may be taken as 372 K, and the equilibrium data are:

Mole fraction of heptane in vapor 0.96 0.91 0.83 0.74 0.65 0.50 0.37 0.24
Mole fraction of heptane in liquid 0.92 0.82 0.69 0.57 0.46 0.32 0.22 0.13

Determine the minimum reflux ratio required. What diameter column would be required if the reflux used were twice the minimum possible?

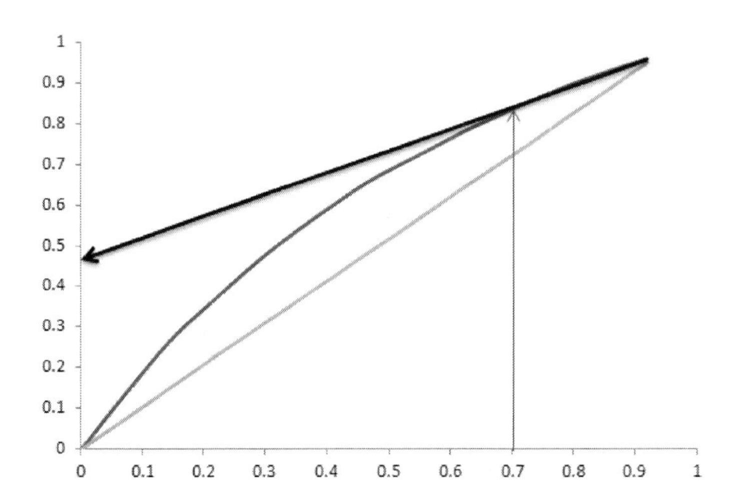

heptane in vapour	0.96	0.91	0.83	0.74	0.65	0.5	0.37	0.24	0
heptane in liquid	0.92	0.82	0.69	0.57	0.46	0.32	0.22	0.13	0

Xd = 0.695

: −0.475 = xD/(Rm + 1) and Rm = 1.08

If 2Rm is used, then: R = 2.16 Ln/D = 2.16
Taking 100 kmol of feed, as a basis, an overall mass balance and a balance for the

100 = (D + W)

and: 100 × 0.695 = 0.99D + 0.01W since 99 per cent n-octane is required.
Hence: D = 69.9 and W = 30.1

and: Ln = 2.16D = 151 and Vn = Ln + D = 221
The mean molecular mass of the feed = (0.695 × 100) + (0.305 × 114) = 104.3 kg/kmole.

Thus: feed rate = (1.25/104.3) = 0.0120 kmole/s
The vapor flow at the top of the column = (221/100) × 0.0120 = 0.0265 kmole/s.
The vapor density at the top of the column = (1/22.4) (273/372) =01

Hence, the volumetric vapor flow = (0.0265/0.0328) = 0.808 m^3/s
If the vapor velocity = 0.6 m/s, the are a required = (0.808/0.6) = 1.35 m^2 equivalent to a column diameter of [(4 × 1.35)/π]0.5 = 1.31 m.

REFERENCE

Microsoft Office Excel 2007, Excel 2007 Win32 English VUP CD, http://excelcalculations.blogspot.com/?, Mindspark Interactive Network, Inc., 2016.

8 Reaction Kinetics, Chemical Reactors, and Thermodynamics

Chapter 8 consists of three parts: Part I "Fundamentals of Reaction Kinetics," Part II "Types and Design of Chemical Reactors," and Part III "Introduction to Thermodynamics."

In Part I, the two forms of the rate law for chemical kinetics are presented: the *differential rate law* and the *integrated rate law*. Most chemical reactions obey one of the three differential rate laws:

1. Zero-order reactions (for which reaction rates are independent of concentration)
2. First-order reactions
3. Second-order reactions

The main factors that influence the reaction rate are discussed. For each integrated rate law, the characteristic plots are shown and presented in table form as well.

A summary of the types of laboratory and industrial catalytic reactors is presented in Part II, along with the equations underlying the design of a chemical reactor.

Part III, on the other hand, introduces thermodynamics. Thermodynamics applies to a wide variety of topics in science and engineering, especially physical chemistry, chemical engineering, thermal power generation, and industrial applications of steam and combustion turbines. It is defined as the subject that relates heat to forces acting between contiguous parts of bodies and the relation of heat to electrical energy.

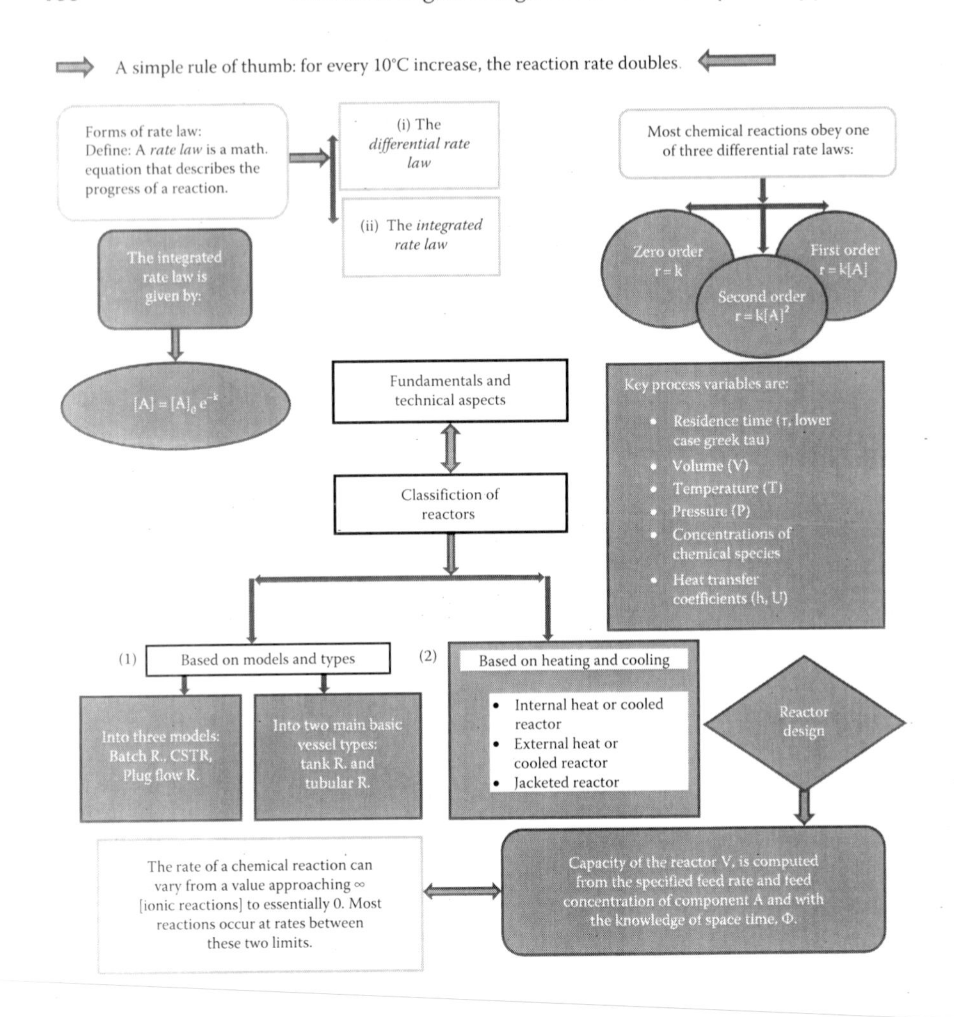

8.1 PART I: FUNDAMENTALS OF REACTION KINETICS

8.1.1 Introduction

Chemical kinetics is the branch of chemistry that addresses the question: "How fast do reactions go?" Chemical kinetics, also known as reaction kinetics, could be described as the study of rates of chemical processes.

In this respect, chemistry can be thought of, at the simplest level, as the science that concerns itself with making new substances from other substances.

If chemistry is making new substances out of *old* substances, through a chemical reaction, then there are two basic questions that confront us and need to be resolved ahead of any action to be taken.

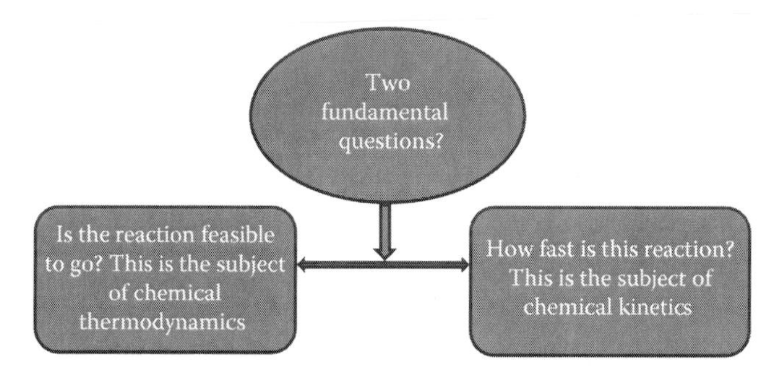

One concludes that thermodynamics gave us a partial answer to our problem. Not only do we have to know whether a reaction is thermodynamically favored, but we also have to know whether the reaction can or will proceed at a finite rate. This completes the other part of the chemistry story by the study of the rate of reactions or chemical kinetics.

8.1.2 Six Categories of Chemical Reactions

According to Myers (1951), all chemical reactions can be placed into one of six categories:

1. *Combustion*: A combustion reaction is when oxygen combines with another compound to form water and carbon dioxide:

$$CH_4 + 2O_2 \rightarrow 2H_2O + CO_2.$$

2. *Synthesis*: A synthesis reaction is when two or more simple compounds combine to form a more complicated one.

   One example of a synthesis reaction is the combination of iron and sulfur to form iron sulfide:

$$8\,Fe + S_8 \rightarrow 8\,FeS$$

3. *Decomposition*: A decomposition reaction is the opposite of a synthesis reaction—wherein a complex molecule breaks down to make simpler ones.

   One example of a decomposition reaction is the electrolysis of water to make oxygen and hydrogen gas:

$$2H_2O \rightarrow 2H_2 + O_2$$

4. *Single displacement*: This is when one element trades places with another element in a compound.

   One example of a single displacement reaction is when magnesium replaces hydrogen in water to make magnesium hydroxide and hydrogen gas:

$$Mg + 2H_2O \rightarrow Mg(OH)_2 + H_2$$

5. *Double displacement*: This is when the anions and cations of two different molecules switch places, forming two entirely different compounds.

   One example of a double displacement reaction is the reaction of lead nitrate with potassium iodide to form lead iodide and potassium nitrate:

$$Pb(NO_3)_2 + 2KI \rightarrow PbI_2 + 2KNO_3$$

6. *Acid–base reaction*: This is a special kind of double displacement reaction that takes place when an acid and base react with each other.

   One example of an acid–base reaction is the reaction of hydrobromic acid (HBr) with sodium hydroxide:

$$HBr + NaOH \rightarrow NaBr + H_2O$$

8.1.3 Reaction Rates

Chemical kinetics deals with the experimental determination of reaction rates from which rate laws and rate constants are derived. In many reactions, the rate of reaction changes as the reaction progresses. Initially the rate of reaction is relatively large, while at a very long time, the rate of reaction decreases to zero (at which point the reaction is complete). In order to characterize the kinetic behavior of a reaction, it is desirable to determine how the rate of reaction varies as the reaction progresses.

The main factors that influence the reaction rate include

1. The physical state of the reactants
2. The concentrations of the reactants
3. The temperature at which the reaction occurs
4. The presence of catalysts

8.1.3.1 Forms of Rate Law

A *rate law* is a mathematical equation that describes the progress of the reaction. In general, rate laws must be determined experimentally. Unless a reaction is an elementary reaction, it is not possible to predict the rate law from the overall chemical equation. There are two forms of a rate law for chemical kinetics:

1. The *differential rate law*
2. The *integrated rate law*

The differential rate law relates the rate of reaction to the concentrations of the various species in the system. Differential rate laws can take on many different forms, especially for complicated chemical reactions. However, most chemical reactions obey one of three differential rate laws:

1. Zero-order reactions (for which reaction rates are independent of concentration)
2. First-order reactions
3. Second-order reactions

Each rate law contains a constant, k, called the *rate constant*. The units for the rate constant depend upon the rate law, because the rate always has units of mole/L · s and the concentration always has units of mole/L.

It should be stated that the rate constant, k, is dependent on the temperature of which the reaction takes place. This can be seen through the Arrhenius equation shown as follows:

$$k = Ae^{-E/RT}$$

In addition to temperature, the rate constant k is dependent on the activation energy, E_a (in Joules). "A" in the equation represents a preexponential factor that has the same units as k. Finally, R is the universal gas constant.

8.1.3.1.1 Zero-Order Reaction

For a zero-order reaction, the rate of reaction is a constant. When the limiting reactant is completely consumed, the reaction stops. The differential rate law is given by the following equation:

$$r = k \tag{8.1}$$

The rate constant, k, has units of mole/L · s.

8.1.3.1.2 First-Order Reaction

For a first-order reaction, the rate of reaction is directly proportional to the concentration of one of the reactants. The differential rate law is given by the following equation:

$$r = k[A] \tag{8.2}$$

The rate constant, k, has units of s^{-1}.

8.1.3.1.3 Second-Order Reaction

For a second-order reaction, the rate of reaction is directly proportional to the square of the concentration of one of the reactants. The differential rate law is given by the following equation:

$$r = k[A]^2 \tag{8.3}$$

The rate constant, k, has units of L/mole · s.

The behavior of these three described rates is illustrated as follows:

1. For a zero-order reaction, the rate of reaction is constant as the reaction progresses.
2. For a first-order reaction, the rate of reaction is directly proportional to the concentration. As the reactant is consumed during the reaction, the concentration drops and so does the rate of reaction.
3. For a second-order reaction, the rate of reaction increases with the square of the concentration, producing an upward curving line in the rate–concentration plot. For this type of reaction, the rate of reaction decreases rapidly (faster than linearly) as the concentration of the reactant decreases.

8.1.3.1.4 Integrated Rate Laws

As was demonstrated, the differential rate law describes how the rate of reaction varies with the concentrations of various species, usually reactants, in the system. The rate of reaction is proportional to the rates of change in concentrations of the reactants and products, that is, the rate is proportional to a derivative of a concentration. This is explained as follows:

For the reaction

$$A \rightarrow B$$

The rate of reaction, r, is simply described by

$$r = -\frac{d[A]}{dt} \tag{8.4}$$

Assuming that this reaction is a first-order, as given by the following equation:

$$r = k[A], \tag{8.2}$$

one can write the rate equation as

$$r = -\frac{d[A]}{dt} = k[A] \tag{8.5}$$

This equation is a differential equation that relates the rate of change in a concentration to the concentration. By multiplying both sides of Equation 8.5 by dt/[A], one obtains the following equation:

$$\frac{d[A]}{[A]} = -k\,dt \tag{8.6}$$

Integration of this equation between the time limits, t = o and t = t, gives the corresponding *integrated rate law* (Equation 8.7), which relates the concentration to time:

$$[A] = [A]_0 e^{-kt} \quad (8.7)$$

where at t = 0, the concentration of A is $[A]_0$.

For each integrated rate law, there is a characteristic plot that can be created that will produce a straight line. These characteristic plots are presented in the following table; species A is a reactant in the chemical reaction (Dixon, 2002).

Reaction Order	Differential Rate Law	Integrated Rate Law	Characteristic Kinetic Plot	Slope of Kinetic Plot	Units of Rate Constant
Zero	$-\frac{d[A]}{dt} = k$	$[A] = [A]_0 - kt$	[A] vs. t	–k	mole/L·s
First	$-\frac{d[A]}{dt} = k[A]$	$[A] = [A]_0 e^{-kt}$	ln [A] vs. t	–k	s^{-1}
Second	$-\frac{d[A]}{dt} = k[A]^2$	$[A] = \frac{[A]_0}{1 + kt[A]_0}$	1/[A] vs. t	k	L/mole·s

The series of three graphs shown later illustrates the use of the characteristic kinetic plots. The graph on the left shows [A] versus t plots for a zero-order (red line), first-order (green line), and second-order (blue line) reaction. The graph in the middle shows ln [A] versus t plots for each reaction order, and the graph on the right shows 1/[A] versus t plots for each reaction order.

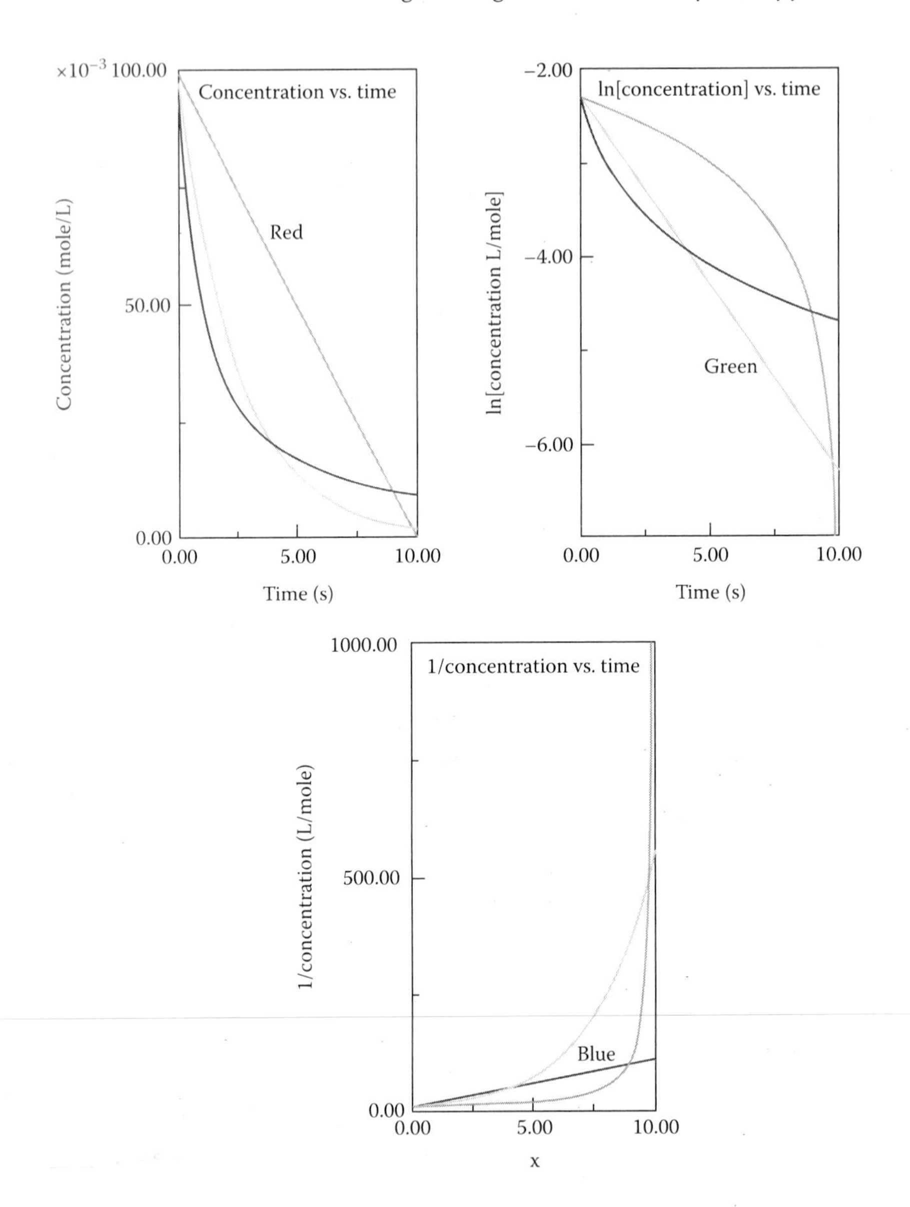

8.2 PART II: TYPES AND DESIGN OF CHEMICAL REACTORS

8.2.1 Introduction

Chemical reactors are generally classified into three main models:

1. Batch reactor model (batch)
2. Continuous stirred-tank reactor model (CSTR)
3. Plug flow reactor model (PFR)

On the other hand, there are two main basic vessel types (in accordance to their shapes):

1. A tank reactor
2. A pipe or tubular reactor (laminar flow reactor)

Both types can be used as continuous reactors or batch reactors, and either may accommodate one or more solids (reagents, catalyst, or inert materials), but the reagents and products are typically fluids.

A chemical reactor, typically a tubular reactor, could be a packed bed. The packing inside the bed may have a catalyst to catalyze the chemical reaction. A chemical reactor may also be a fluidized bed.

Furthermore, catalytic reactors require separate treatment, whether they are batch, CST, or PF reactors, as the many assumptions of the simpler models are not valid.

8.2.2 Types of Laboratory and Industrial Catalytic Reactors

Laboratory catalytic reactors include many types: pulse, batch, tubular (differential and integral), CSTR, recycle, and transport. Figure 8.1 is an illustration of this group. The details of CSTR are given in Figure 8.2.

Industrial catalytic reactors, on the other hand, cover different types such as fixed bed, trickle bed, moving bed, fluidized bed, slurry, and monolith. This set of reactors is schematically shown in Figure 8.3. The details of a fixed-bed reactor are shown in Figure 8.4.

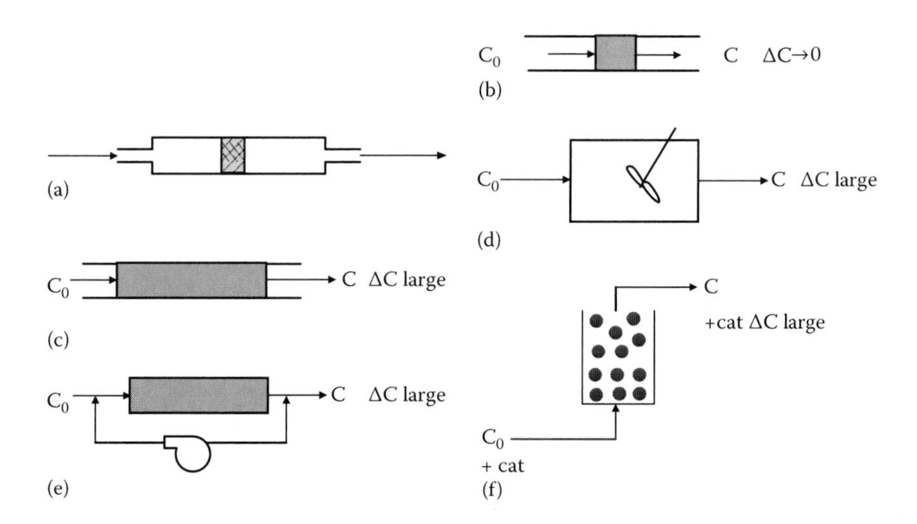

FIGURE 8.1 Types of laboratory catalytic reactors: (a) pulse, (b) differential, (c) integral, (d) CSTR, (e) recycle, and (f) transport.

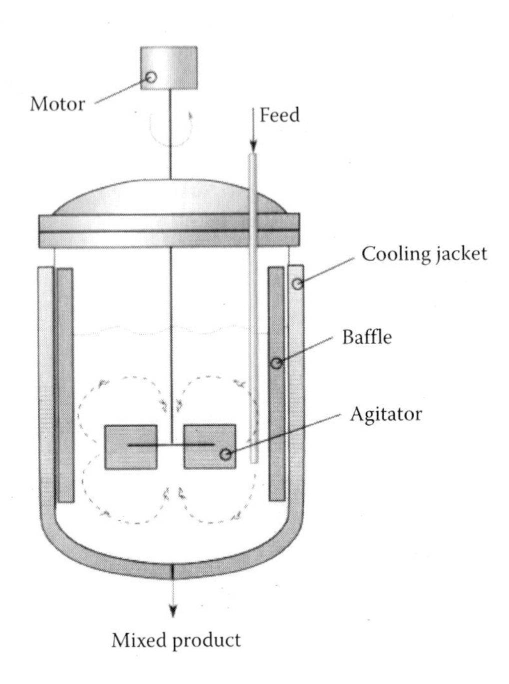

FIGURE 8.2 CSTR.

8.2.3 Catalytic Reactors

In a catalytic reactor, the rate of a catalytic reaction is proportional to

1. The effective amount of catalyst
2. The concentration of the reactants

With a solid phase catalyst and fluid phase reagents, this is proportional to the exposed area, efficiency of diffusion of reagents in and products out, and efficacy of mixing.

The behavior of the catalyst is also a consideration. Particularly in high-temperature petrochemical processes, catalysts are deactivated by sintering, coking, and similar processes.

A common example of a catalytic reactor is the catalytic converter following an engine. However, most petrochemical reactors are catalytic and are responsible for most of industrial chemical production in the world, with extremely high-volume examples such as sulfuric acid, ammonia, and reformate/benzene, toluene, ethylbenzene, and xylene (BTEX).

A catalyst performance depends directly on the following four factors:

1. Activity
2. Selectivity
3. Life and stability
4. Economics and process costs

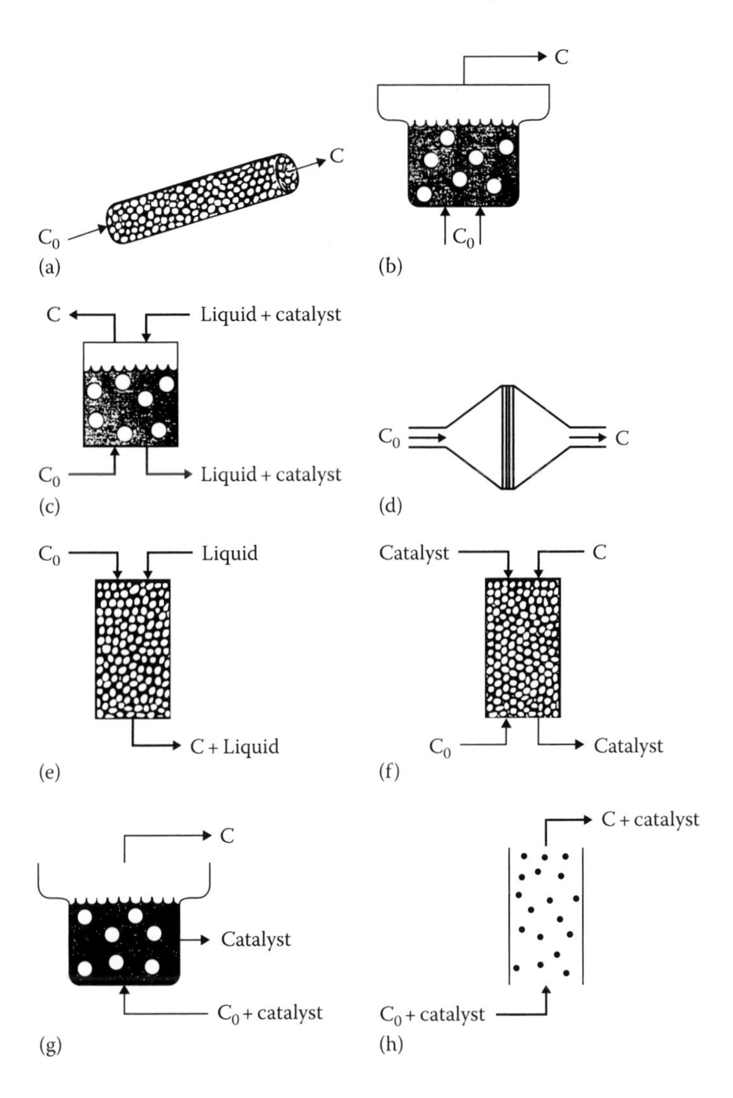

FIGURE 8.3 Heterogeneous catalysis reactor types: (a) fixed bed, (b) batch fluid bed, (c) slurry, (d) catalytic gauze, (e) trickle bed, (f) moving bed, (g) continuous fluid bed, and (h) transport line.

8.2.4 Design of Chemical Reactors

Reactors are designed based on features like

1. Mode of operation, *batch versus continuous*
2. Types of phases present, *homogeneous versus heterogeneous*
3. Geometry of reactors, *vessel versus tube*

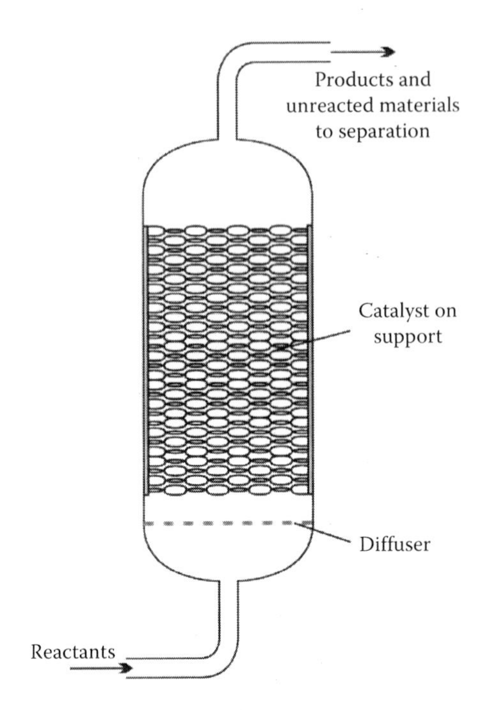

FIGURE 8.4 Fixed-bed reactor (e.g., this is used in the manufacture of ammonia).

Key process variables in chemical reaction engineering include

- Residence time (θ)
- Volume (V)
- Temperature (T)
- Pressure (P)
- Concentrations of chemical reactants (C_A…)

8.2.4.1 Basic Design Equations

The parameter that has been commonly used as a measure of the reactor capacity is either

- The mean residence time θ_m
- The space time φ

Definition of θ_m: It is the average of time periods during which reaction mixtures stay in the reactor and is given by the equation

$$\theta_m = \int \frac{dN_A}{V(-r_A)}, \quad \text{between the limits } N_{A0} \text{ and } N_{Af} \tag{8.8}$$

where

N_A is the number of moles of reactant A
V is the reactor volume
$-r_A$ is the rate of disappearance of A

In terms of the fractional conversion, x_A, Equation 8.8 becomes

$$\theta_m = N_{A0} \int \frac{dx_A}{V(-r_A)}, \quad \text{between the limits 0 and } x_{Af} \tag{8.9}$$

Under steady-flow conditions with constant fluid density, θ_m for a continuous flow reactor may be calculated by using

$$\theta_m = \frac{V}{q} \tag{8.10}$$

where q is the volumetric flow rate of the reaction mixture.

Definition of ϕ, *the space time*: It is defined as the time elapsed in processing one reactor volume of feed at specified conditions. For example, if a value of ϕ is 1.5 h, it implies that 1.5 h would be required to process a one rector volume of feed at known conditions. It is given by the equation

$$\phi = \frac{V}{(F/\ell)} = \frac{V}{(F_{A0}/C_{A0})} = \frac{V}{q_0} = \frac{1}{S} \tag{8.11}$$

Sizing of a reactor may be illustrated by a general example for *a PF* reactor as illustrated in the following steps:

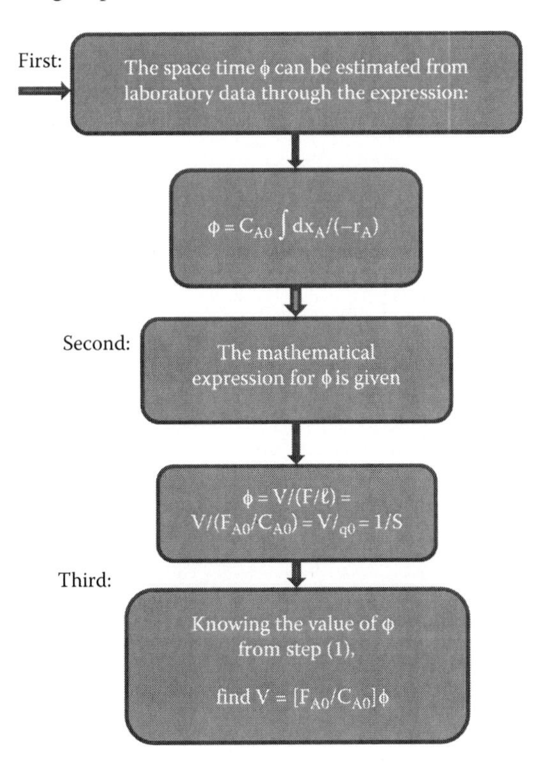

where

- F and F_{A0} are the flow rates in moles per unit time of the total feed and of reactant A in the feed, respectively
- ℓ is the molar density in moles per unit volume of the feed
- C_{A0} is the concentration of A in the feed
- q_0 is the volumetric flow rate of the feed at entering conditions
- S is the space velocity

To summarize this, the required capacity of the reactor V is computed from the specified feed rate and feed concentration of component A and with the knowledge of space time, ϕ.

8.3 PART III: INTRODUCTION TO THERMODYNAMICS

8.3.1 Basic Definitions and Concepts

To start with, thermodynamics could be described as the study of energy conversion, most typically through terms of heat and work.

8.3.1.1 The Concept of Equilibrium

The state of a system in which properties have definite, unchanged values as long as external conditions are unchanged is called an equilibrium state (Figure 8.5).

A system in thermodynamic equilibrium satisfies

- Mechanical equilibrium (no unbalanced forces)
- Thermal equilibrium (no temperature differences)
- Chemical equilibrium

8.3.1.2 The Concept of a Process

If the state of a system changes, then it is said that *it is undergoing a process.* The succession of states through which the system passes defines the path of the process.

8.3.1.3 Equations of State

It is an experimental fact that two properties are needed to define the state of any pure substance in equilibrium or undergoing a steady-steady process.

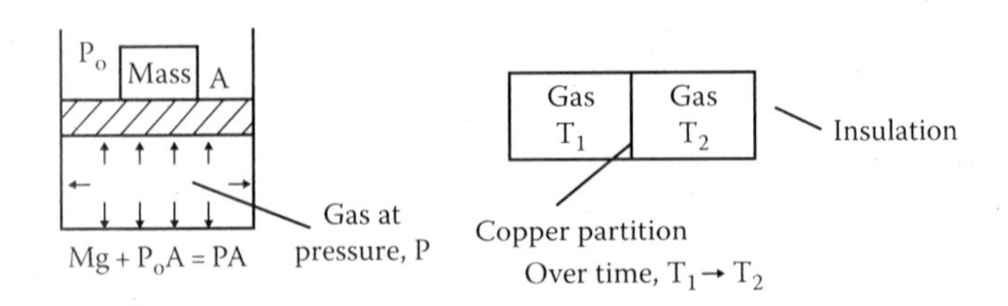

FIGURE 8.5 Illustration of mechanic and thermodynamic equilibrium.

Thus, for a simple compressible gas like air,

$$P = P(\upsilon, T), \quad \text{or} \quad \upsilon = \upsilon(P, T), \quad \text{or} \quad T = T(P, \upsilon),$$

where υ is the volume per unit mass, $(1/\ell)$, that is, if we know υ and T is the temperature, we know P is the pressure, etc. Any of these is equivalent to the equation $f(P, \upsilon, T) = 0$, which is known as an equation of state. The equation of state for an ideal gas, which is a very good approximation to real gases at conditions that are typically of interest for aerospace applications, is

$$P\bar{\upsilon} = \mathcal{R}T,$$

where

$\bar{\upsilon}$ is the volume per mol of gas

$\mathcal{R}$ is the *universal gas constant*

8.3.2 Thermodynamic Laws and Entropy

- *Zeroth law*: When two systems are both in *thermal equilibrium* with a third system, then they must be in thermal equilibrium with each other, that is, thermal equilibrium is transitive.
  If Temp(A) = Temp(B) and Temp(B) = Temp(C), then Temp(A) = Temp(C).
- *First law*: *The increase in internal energy of a closed system is equal to the difference of the heat supplied to the system and the work done by it:* $\Delta U = Q - W$.
  Note that due to the ambiguity of what constitutes positive work, some sources state that $\Delta U = Q + W$, in which case work done on the system is positive.
- *Second law*: The second law states that all work processes tend toward greater entropy over time. Another way of saying this is that total entropy of an isolated system can never decrease.
  In other words, heat cannot spontaneously flow from a colder location to a hotter location.
- *Third law*: *As a system approaches absolute zero, the entropy of the system approaches a minimum value.*
  The third law of thermodynamics is a statistical law of nature regarding entropy and the impossibility of reaching absolute zero of temperature.

8.3.2.1 Entropy

The idea of *entropy* comes from a principle of thermodynamics dealing with energy. It usually refers to the idea that everything in the universe eventually moves from order to disorder and *entropy* is the measurement of that change. The following is a list of definitions of entropy from a collection of textbooks:

- A measure of energy dispersal at a specific temperature
- A measure of disorder in the universe or of the availability of the energy in a system to do work
- A measure of a system's thermal energy per unit temperature that is unavailable for doing useful work

We see evidence that the universe tends toward highest entropy many places in our lives. A campfire is an example of entropy. The solid wood burns and becomes ash, smoke, and gases, all of which are more disordered than the solid fuel. For isolated systems, entropy never decreases. This fact has several important consequences in science: it prohibits *perpetual motion* machines, it implies that the arrow of entropy has the same direction as the arrow of time, and so on. Increases in entropy corresponds to irreversible changes in a system, because some energy is expended as waste heat, limiting the amount of work a system can do.

8.3.3 Role of ΔG in Chemical Reactions

For the general chemical reaction

$$aA + bB \underset{r_2}{\overset{r_1}{\rightleftarrows}} cC$$

The forward rate of the reaction is

$$r_1 = k_1[A]^a[B]^b$$

The backward rate is

$$r_2 = k_2[C]^c$$

Now, at equilibrium

$$r_1 = r_2$$

Therefore,

$$k_1[A]^a[B]^b = k_2[C]^c$$

and we get

$$\frac{k_1}{k_2} = K_p = \frac{[C]^c}{[A]^a[B]^b} = \frac{[\text{Products}]}{[\text{Reactants}]}$$

Given the free energy of formation of the reaction, ΔG, we can write

$$\Delta G = -RT \ln K_p$$

where K_p is the equilibrium constant for the reaction at the given conditions of T and P, expressed as the concentrations of the participating species in terms of the partial pressures.

ΔG at constant temperature is also given as a function of entropy change:

$$\Delta G = \Delta H - T\Delta S$$

where ΔH and ΔS are the increase in total heat content and entropy in the reaction, respectively.

END-OF-CHAPTER SOLVED EXAMPLES

Part (a): Manual Calculations

Example 8.1

Calculate the entropy of the surroundings for the following two reactions:

(a) $C_2H_8(g) + 5O_2(g) \rightarrow 3CO_2(g) + 4H_2O(g)$

$$\Delta H = -2045\ kJ$$

(b) $H_2O(l) \rightarrow H_2O(g)$

$$\Delta H = +44\ kJ$$

Solution

The change in entropy of the surroundings after a chemical reaction at constant pressure and temperature can be expressed by the formula

$$\Delta S_{surr} = -\frac{\Delta H}{T}$$

where

ΔS_{surr} is the change in entropy of the surroundings
−ΔH is the heat of reaction
T is the absolute temperature in Kelvin

Reaction (a)

$$\Delta S_{surr} = -\frac{\Delta H}{T}$$

$$\Delta S_{surr} = -\frac{-2045\ kJ}{25+273}$$

Remember to convert °C to K

$$\Delta S_{surr} = \frac{2045\ kJ}{298\ K}$$

$$\Delta S_{surr} = 6.86\ kJ/K \quad \text{or} \quad \underline{6860\ J/K}$$

Note that the increase in the surrounding entropy since the reaction was exothermic.

Reaction (b)

$$\Delta S_{surr} = -\frac{\Delta H}{T}$$

$$\Delta S_{surr} = -\frac{+44\ kJ}{298\ K}$$

$$\Delta S_{surr} = -0.15\ kJ/K \quad \text{or} \quad \underline{-150\ J/K}$$

This reaction needed energy from the surroundings to proceed and reduced the entropy of the surroundings.

Conclusions: The change in entropy of the surroundings of reactions 1 and 2 was 6860 and −150 J/K, respectively.

Example 8.2

The following example illustrates on how to determine the thermodynamic feasibility of a chemical reaction:

$$2\ H_2(g) + O_2(g) \rightarrow 2H_2O(l)$$

We can calculate $\Delta_r G°$ for this reaction from the tables of free energies of formation. We find that $\Delta_r G°$ for this reaction is very large and of negative value, which means that the reaction is very feasible. We conclude, right away, that the equilibrium constant, K, for the reaction is very large:

$$\Delta G = -RT \ln K$$

However, we can mix hydrogen gas and oxygen gas together in a container, even in their correct stoichiometric proportions, and they will stay there for centuries, perhaps even forever, without reacting. The problem is not that the reactants do not want to form the products, they do, but they cannot find a *pathway* to get from reactants to products. In other words, thermodynamics is not the only main focus in chemistry. Not only do we have to know whether a reaction is thermodynamically favored, we also have to know whether the reaction can or will proceed at a finite rate. The study of the rate of reactions is called chemical kinetics.

Example 8.3

Methanol vapor can be converted into formaldehyde by the following reaction scheme:

$$CH_3OH + 0.5O_2 \rightarrow HCHO + H_2O$$

$$CH_3OH \rightarrow HCHO + H_2$$

The fresh feed to the process was 0.5 kmole/h of O_2 and an excess methanol. All of the O_2 reacts in the reactor. Formaldehyde and water are removed from the product stream first, after which H_2 is removed from the recycled methanol. The recycle flow rate of methanol was 1 kmole/h. The ratio of methanol reacting by decomposition to that by oxidation was 3. Draw the flow diagram and then calculate the per pass conversion of methanol in the reactor and the fresh feed rate of methanol.

Calculations:
The flow diagram for the aforementioned process is given here:

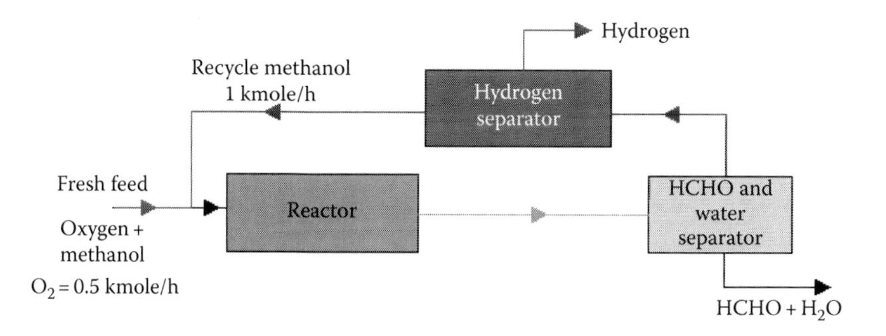

From the given statements, we apply the following:

- All of the O_2 entering reacts in the reactor.
- The ratio of methanol reacting by decomposition to that by oxidation was 3.

The numbers of moles taking part in the reactions are given as follows:

$$\underset{1}{CH_3OH} + \underset{0.5}{0.5\ O_2} \rightarrow \underset{1}{HCHO} + \underset{1}{H_2O} \quad (8.3.1)$$

$$\underset{1}{CH_3OH} + \rightarrow \underset{1}{HCHO} + \underset{1}{H_2} \quad (8.3.2)$$

From reaction (8.3.1): In 1 h, 0.5 mole of O_2 reacts with 1 mole of CH_3OH.

Therefore, for complete conversion of oxygen, 0.5 moles/h of O_2 needs 1 mole/h of CH_3OH.

Obtaining the ratio of methanol conversion, we then find that the flow rate in methanol decomposition is three times greater than that in methanol oxidation, that is, 3 moles/h of CH_3OH.

Therefore,

Total methanol entering the reactor = methanol reacted by oxidation + methanol reacted by decomposition + recycled methanol = 1 + 3 + 1 = 5 kmoles/h

Conversion per pass = 100 × (total methanol reacted/total methanol entering the reactor) = 100 × 4/5 = 80%

Fresh methanol rate = total methanol entering the reactor − recycled methanol = 5 − 1 = 4 kmoles/h (Subbu, 2014)

Part (b): Using MATLAB®

Example 8.4: Chemical Equilibrium Inside a Chemical Reactor

Methanol is formed by reacting carbon monoxide with hydrogen. (This might be an onboard generation of hydrogen for the hydrogen economy.) With the reaction

$$CO + 2H_2 \Leftrightarrow CH_3OH,$$

the condition at equilibrium is

$$Kp = \frac{p_{CH_3OH}}{pCOpH_2^{\,2}} = \frac{y_{CH_3OH}}{yCOyH_2^{\,2}} \frac{1}{P^2}$$

When the carbon monoxide and hydrogen enter the reactor in a 1: 2 ratio, find the equilibrium conversion when the pressure is 50 atm and Kp = 0.0016.

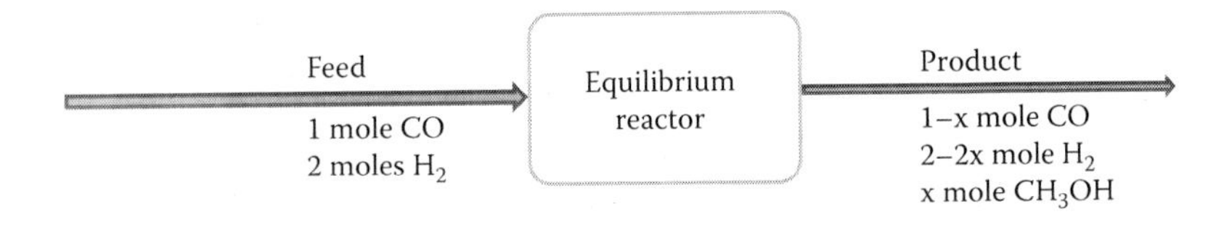

A conceptual model is formed as follows, where x is the fraction reacted from the carbon monoxide at equilibrium.
Hence,

$$Kp = 0.0016 = \frac{x(3-2x)^2}{(1-x)(1-x)^2} \frac{1}{4 \times 50^2}$$

By solving this equation for x, the equilibrium conversion is found to be 0.50. Solution could be obtained using Matlab,

Example 8.5

Methane is used to produce acetylene as the main product by the following thermal reaction:

$$2CH_4(gas) + Q \rightarrow C_2H_2(g) + H_2(g)$$

As an undesired side reaction, acetylene decomposes to give

$$C_2H_2 \rightarrow 2C(s) + H_2(g)$$

Given

- Methane is fed to the reactor at 1500°C, at a rate of 10.0 moles/s
- Heat is transferred to the reacting system at the rate of 975 kW
- The product temperature is 1500°C
- The % fractional conversion of methane is 60
- The enthalpy balance $Q = \Delta H = \left[\sum n_i H_i\right]_{products} \left[\sum n_i H\right]_{reactants}$
- $H_i = \left(\Delta H_f^0\right)_I + C_{pi}(1500 - 25)$

Calculate the product component flow rates and the yield of moles of C_2H_2/mole of CH_4. Also, calculate the yield of H_2 per mole of H_2 introduced to the reactor in the form of CH_4.

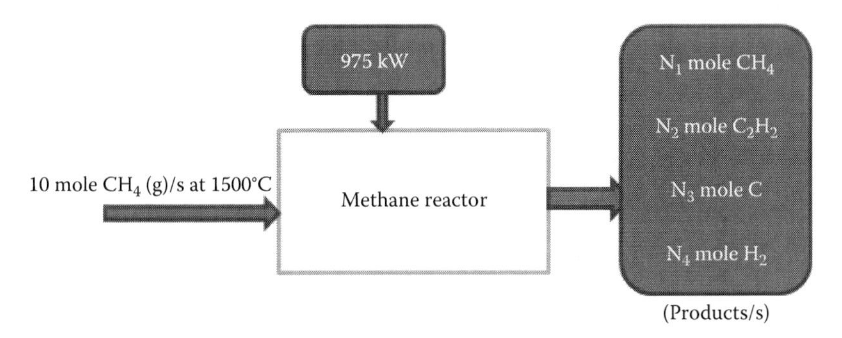

Solution

The conversion rate of methane *is 60%*; hence, $N_1 = 10(1 - 0.6) = 4$ moles CH_4 (*unreacted*).

Carbon balance, in = out

$$10(1) = 4(1) + 2(N_2) + N_3$$

or

$$2N_2 + N_3 = 6 \tag{8.5.1}$$

Hydrogen balance, in = out

$$10(4) = 4(4) + 2N_2 + 2N_4$$

or

$$2N_2 + 2N_4 = 24 \tag{8.5.2}$$

Note that the moles of hydrogen in this balance are meant to be *atoms* of hydrogen.

Now, we have two equations and three unknowns (N_2, N_3, N_4). We use the enthalpy balance next to come up with the third relationship, in order to have three equations in three unknowns.

$$Q = \Delta H = \left[\sum n_i H_i\right]_{products} - \left[\sum n_i H\right]_{reactants}$$

$$975 = [4H_1 + N_2H_2 + N_3H_3 + N_4H_4] - 10H_1$$

Rearranging and substituting for the enthalpies, Hs with the corresponding values, we obtain

$$303.45\ N_2 + 32.45\ N_3 + 45.72\ N_4 = 1225.08 \qquad (8.5.3)$$

Now, we can enter these three equations, 8.5.1 through 8.5.3, into the *MATLAB matrix form*: A = xb. Solving for the value of

$$x = 2.4983;\ \text{that is } N_2,\ \text{mole of } C_2H_2/s$$

$$= 1.0035;\ \text{that is } N_3,\ \text{mole of C/s}$$

$$= 9.5017;\ \text{that is } N_4,\ \text{mole of } H_2/s$$

Yield of acetylene = 2.4983/6 = 0.417 mole/mole of methane

$$\text{Yield of hydrogen} = 9.5017/\left[10\ \text{moles methane} \times 4\left(\text{hydrogen atoms per mole}\right)\right]$$
$$= 0.2375\ \text{mole/mole of methane}$$

REFERENCES

Dixon, J., Espenson, J. H., Kinetics and mechanism of oxygen atom abstraction from a dioxo-rhenium(VII) complex, *Inorganic Chemistry* 41: 4727, 2002.

Myers, R. *The Basics of Chemistry*, London, U.K.: Greenwood Publishing Group, 1951.

Subbu, M. *Chemical Engineering Learning Resources*, 2014. http://www.msubbu.in.

9 Chemical Plant Design and Process Economics

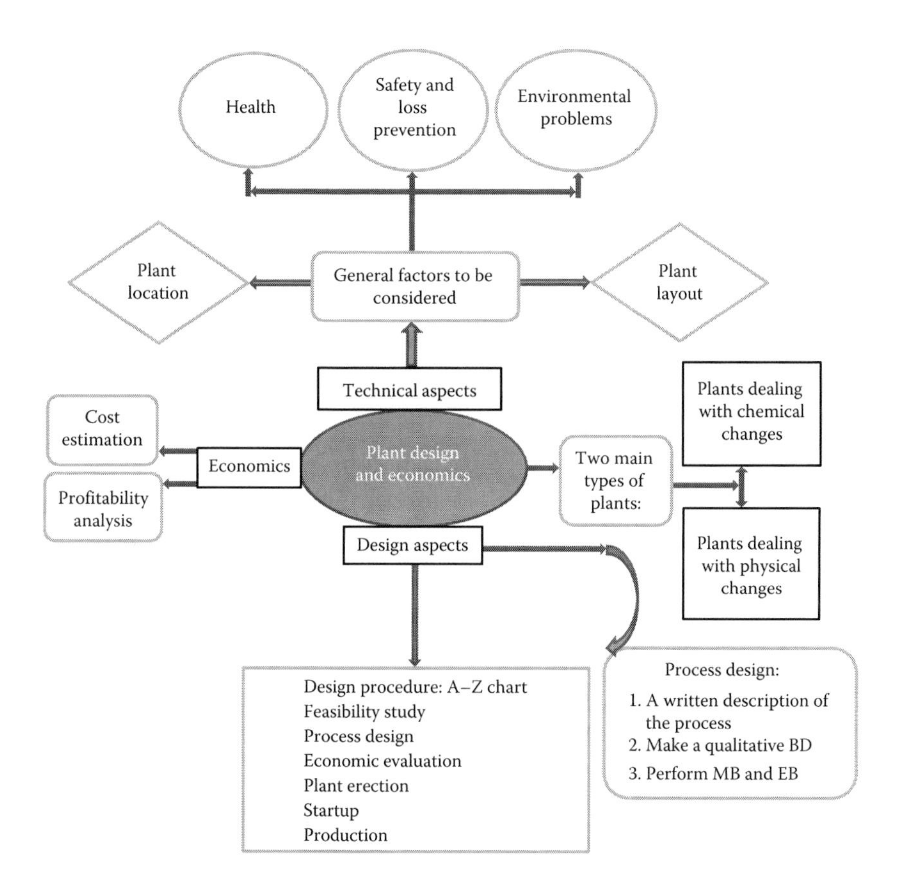

The successful convergence of theory and practice has a great impact on developing courses in plant design. This chapter will highlight the different aspects involved in the design of chemical plants. Since the topic is too involved, we attempt to provide a balanced presentation to the strategy of process synthesis in plant design consistent with industrial practice and within the economic frame of feasible projects. The dollar sign must always be kept in mind when carrying out the design of a chemical plant.

This chapter covers the basic fundamentals in plant design that the reader should acquire, from the inception of an idea all the way to the construction of a plant. Process design is a cornerstone in this presentation. General design considerations, complementary to the design of a plant, are discussed. These include safety, health hazards, and design aspects related to the plant location and plant layout. The economic aspects involved in building a chemical plant, from cost estimation to profitability studies, are presented. Illustrations of the use of computer-aided design (CAD) to study the performance of process units in chemical plants and their integration into a complete plant are presented. Solved examples and sample problems are used extensively in this chapter to illustrate the applications of principles to practical situations.

9.1 INTRODUCTION

A chemical plant is one that belongs to the *chemical industry*. Most processes in the chemical industry involve a *chemical change*, which could imply two distinctive groups of industry:

1. One The first group of industry involves a chemical change, for example, in the case of chemical reaction of nitrogen with hydrogen that produces ammonia.
2. The second group of industry involves physicochemical change, involved in the separation of a given feed to components or fractions, for example, the distillation of hydrocarbons and water desalination. These industries are normally recognized as *physical separation operations* that utilize strictly unit operations.
   Purely mechanical changes or separations are not considered part of the chemical processing.

Plant design for processes of the first group utilizes, in its initial stage, the basic chemical stoichiometric equations describing the chemical reactions, in order to do material balance (MB) and heat balance (HB) calculations. However, for industries of the second group, calculations are usually based on setting up the total MB and component MB, for example, the solution of binary-distillation problems involved in the setting up of two equations in two unknowns, as was presented in Chapter 6.

9.2 PLANT DESIGN IS THE HEART OF CHEMICAL ENGINEERING

Modern chemical engineers are concerned with processes that convert raw materials or chemicals into more useful or valuable forms. They are also concerned with the pioneering valuable materials and related techniques—which are often essential to related fields such as nanotechnology, fuel cells, and bioengineering.

Students study many basic courses in chemical engineering that include chemical reaction engineering, heat transfer, and thermodynamics. However, when it comes to plant design, simulated by a heart as shown in Figure 9.1, all of these courses along with other information represent the input and the output design, which is a profitable working entity.

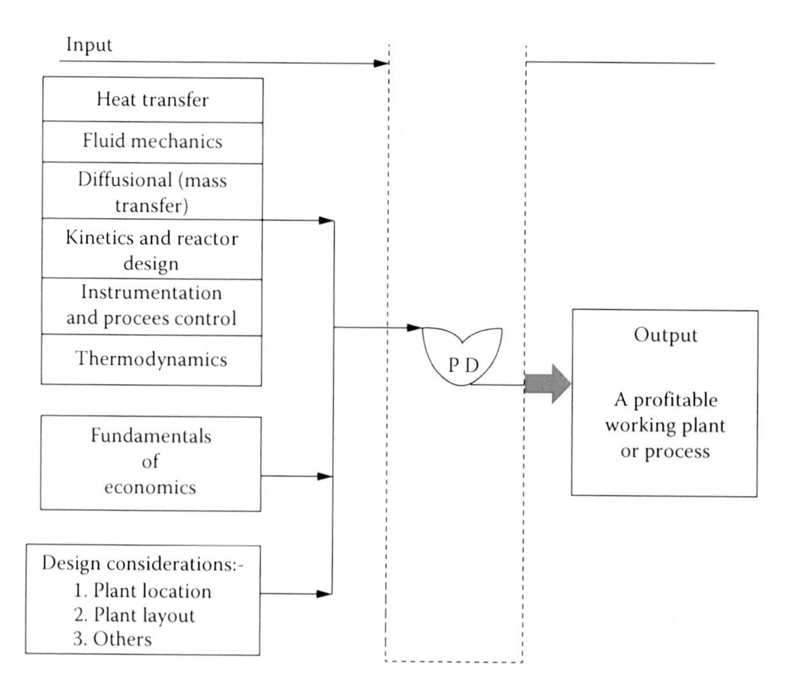

FIGURE 9.1 Plant design is the heart of chemical engineering.

In a plant design course, students are basically well prepared to launch a design project, once they cover the following additional topics: economics, materials selection, computer-aided design (CAD), and optimization technique, and others. Their personal information has an impact in handling a design project as well.

9.3 HOW TO HANDLE A DESIGN PROJECT?

9.3.1 Proposed Procedure

The design of a chemical plant would normally go through the following steps:

- Derive an idea (e.g., to produce a product).
- Find out if it is feasible to build a plant (technical and economic feasibility study).
- Carry out a *process design* that involves three basic stages:
    - Draw a *qualitative* block diagram based on a written description for the selected process.
    - Carry out basic calculations using MB and EB to come up with a quantitative block diagram. MB is the basis of process design, as shown in Figure 9.2.
    - Determine the size and capacity of equipment (equipment sizing).
- Do cost estimation for the capital investment of the plant.
- Carry out profitability analysis for the project.

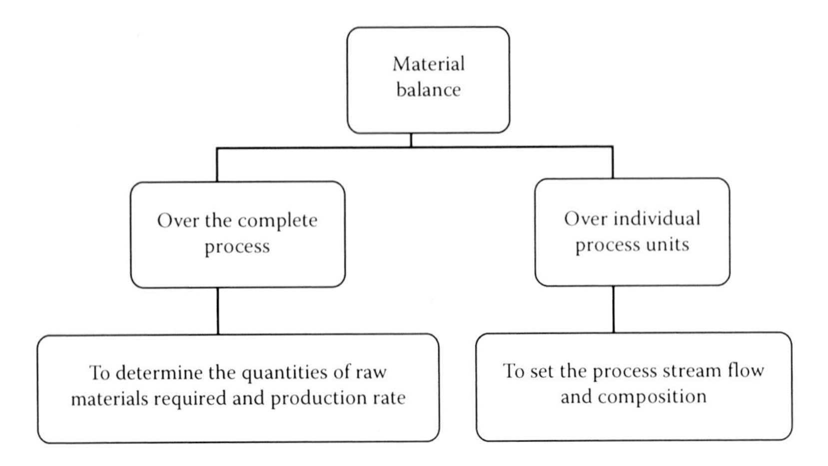

FIGURE 9.2 The role of material balance in process design.

When it comes to computer applications, spreadsheet software has become an indispensable tool in plant design because of the availability of personal computers, ease of use, and adaptability to many types of problems. On the other hand, many programs are available for the design of individual units of chemical process units. The Computer Aids for Chemical Engineering Education Corporation makes available several programs mainly for educational use.

9.3.2 A–Z Chart for Plant Design

In order to make the picture clearer, all steps to be carried out in a design project are put together in the form of what is called A–Z chart, shown in Figure 9.3. It is

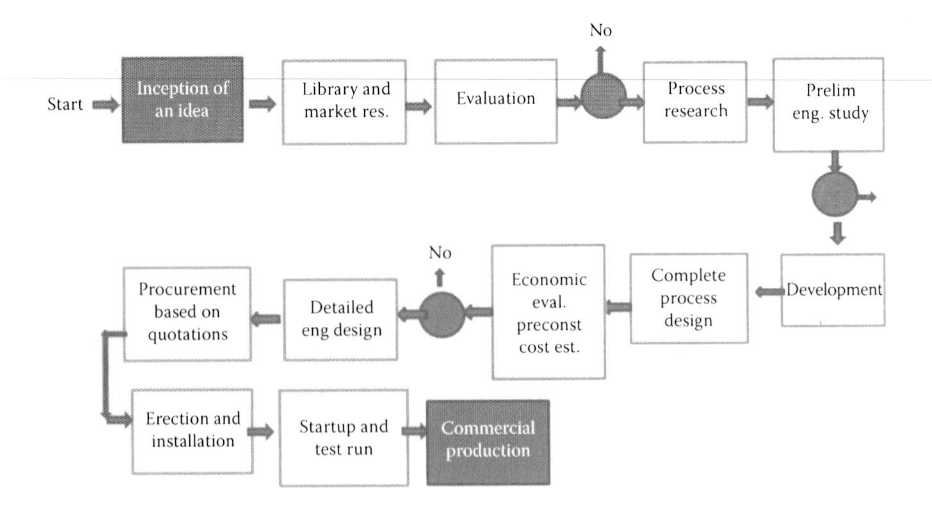

FIGURE 9.3 The A–Z chart in plant design.

a collective and comprehensive flowchart that takes the student through the whole plant to be: from feasibility study to commercial production via the intermediate steps of process design followed by economic evaluation.

In general, the following three cases of design projects are common to handle:

1. New process development
2. New production capacity to meet growing sales
3. Modification and addition to existing plant

9.3.3 Flow Sheeting and Types of Flow Diagrams

Flow sheeting, as used in CAD, means performing on a specified flow sheet, the calculations necessary to simulate the behavior of the process. These calculations include mass and energy balances.

Process design normally starts with a process scheme or flow sheet. A flow diagram is the *road map* of a given process. It is the key document in process design. It is a diagrammatic model of the process, which describes the process steps in a proper sequence using symbols to represent the various components (equipment, lines, and control instrumentation) that make up the unit. During the plant start-up and subsequent operation, the flow sheet forms a basis for comparison of operating performance with design. It is also used by operating personnel for the preparation of operating manual and operator training.

9.3.4 Scale-Up Practice and Safety Factors

Stages in the development in plant design follow the route, right after laboratory-scale experiments, all the way to full-size commercial production, passing by bench scale and pilot plant tests. These stages are illustrated in Figure 9.4.

Safety factors play an important role when dealing with a design project. It is defined as the amount of *overdesign* in order to account for changes in operating performance of equipment or a vessel. A handout summary is illustrated in Figure 9.5.

9.4 OTHER ASPECTS IN PLANT DESIGN

Apart from the engineering principles considered in the plant design, there are other important functions and items to be considered regarding safety, health, loss prevention, plant location, plant layout, and others.

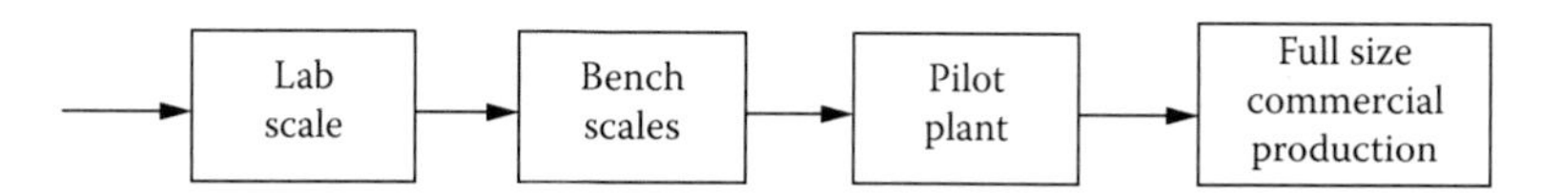

FIGURE 9.4 Stages in plant design.

(i) Definition: It is the amount of ***overdesign*** to account for changes in operating performance of an equipment.

(ii) Applications: 1. To safeguard against ***fouling*** in: (a) heat exchangers and (b) furnace pipes

2. For potential increase in capacity requirements

(iii) Calculations for Ht Ex:

1. To find out if a given Ht Ex is fouled or not?

$Q = U_o A \, \Delta T_m$, $U_o = 1/R_T$

R_T = Total resistance = $R_{fi} + R_w + R_{f0} + R_{si} + R_{so}$, define $R_{si} + R_{so} = R_{st}$

Steps:

- Calculate $U_o = Q/A \, \Delta T_m$
- Find the values of R_{fi}, R_{fo}, and R_w
- Calculate R_{st}, using $U_o = 1/[R_{fi} + R_{fo} + R_w + R_{st}]$
- Compare the value of R_{st}, calculated, with the value for clean Ht Ex....If R_{st} is >> 0.001, then the tube are ***fouled***.

2. Safety factor for overdesign

It is estimated to be about 15%. In other words, increase the calculated area, A, by **15%** as a factor of safety.

FIGURE 9.5 Safety factors in design.

A brief summary is given as follows:

- *Health and safety hazards*: One should consider the toxicity of materials and frequency of exposure, fire, and explosion hazards.
- *Loss prevention*: HAZOP study.
- Environmental protection and control include air pollution, water pollution, solid wastes, thermal effects, and noise effects.
- *For plant location*: Both primary and specific factors are to be considered.
- *For plant layout*: Optimum arrangement of equipment within a given area is a strategic factor.
- *For plant operation and control*: The designer should be aware of
  - Instrumentation
  - Maintenance
  - Utilities
  - Structural design
  - Storage
  - Materials handling, pipes, and pumps
  - Patents aspects

9.5 THE ROLE OF ECONOMICS IN PLANT DESIGN

9.5.1 Introduction

In this chapter, economic fundamentals as being applied to the design of chemical plants are presented. This will involve two main tasks, as explained next:

The steps underlying both categories of cost estimation and profitability analysis are indicated in Figure 9.6.

9.5.2 Estimation of the Fixed Capital Investment and the Total Capital Investment

Obviously, the estimation of the fixed capital investment (FCI) would enable the designer to answer the following questions:

- How much capital do we need for a project?
- What are the annual depreciation costs of the physical assets of a plant?
- How much is the annual maintenance and repair costs of the equipment?
- How much is the working capital (WC)?

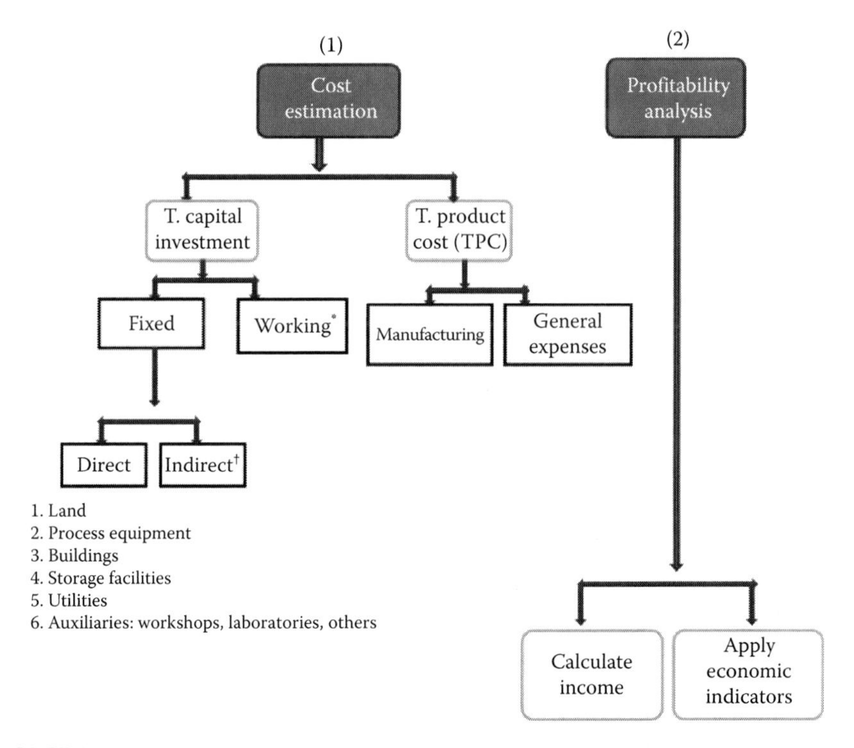

FIGURE 9.6 Economic aspects in plant design. *Working capital is the money needed for raw materials and cash at start up and is equal to 10%–20% of T.C.I. †Indirect fixed capital is spent in engineering expenses, supervision for construction and construction tools and equipment.

Lang factors are recommended in this regard. Once the costs of delivered equipment are estimated, factors are applied as shown next:

	Factor to Estimate	
Classification of the Plant Function	**FCI**	**TCI**
Solid processing operation	3.9	4.6
Solid/fluid processing operation	4.1	4.9
Fluid processing operation	4.8	5.7

The six-tenth factor is another handy method to determine the cost of an equipment of a given capacity (A), if the cost and the capacity of a similar one (B) are known:

$$\text{Cost of A} = \text{Cost of B} \times [\text{Capacity of A}/\text{Capacity of B}]^{0.6}$$

9.5.3 Decision Making Use of Annuity

An annuity is a series of equal payment paid at the end of the year, "A," \$/year. The amount of an annuity, at the end of n years, will be equal to all payments plus interest that accumulate, where i is the interest rate.

The future worth of this sum of money = $A\{(1 + i)^n - 1\}/i = A(SFF)$, where the term $\{(1 + i)^n - 1\}/i$ is known as the *sinking-fund factor* (SFF).

Now, we have two cases:

1. If "A" deposited annually in this sinking fund is the annual depreciation cost of an equipment, A_d, then the future worth of this A_d would simply accumulate to P; the present value of the equipment:

$$P = A_d(SFF) \quad \text{and} \quad A_d = P\left[\frac{i}{(1+i)^n - 1}\right]$$

2. If "A" deposited annually, on the other hand, is the amount of the annual capital recovery of the equipment, A_r, then the amount of money generated by the sinking fund will be the future value for the cost of this equipment: $F = A_r(SFF)$

To make it clearer, the A_d, defined as the annual depreciation cost, when deposited in the sinking fund, will generate the P of the equipment. This is simply to recover the original cost paid in the equipment. On the other hand, when $A_r(>A_d)$, defined as the annual capital recovery, is deposited in the sinking fund, then the amount generated will be F.

In this case, F will be equal to P + the interest on P, as if it is deposited in a bank for n year or simply

$$F = P(1+i)^n$$

This discussion is presented in a concise form as shown in Figure 9.7.

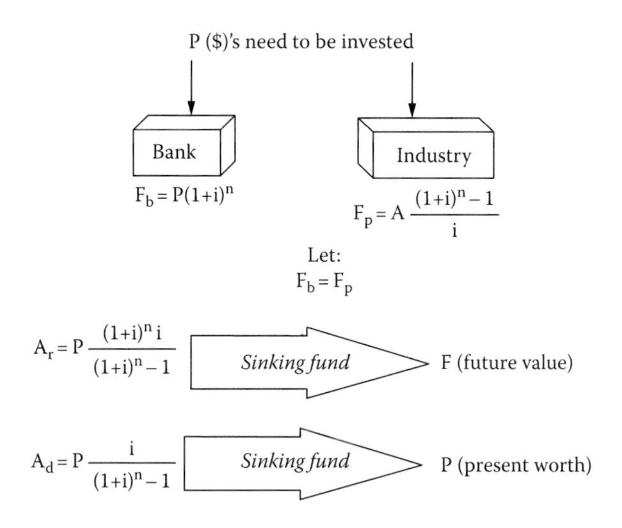

FIGURE 9.7 Investments in bank (F_b) versus industry (F_p).

9.5.4 Profitability Analysis Using Economic Indicators

The basic aim of financial measures and profitability analysis is to provide some yardsticks for the attractiveness of a venture or a project, where the expected benefits (revenues) must exceed the total production costs.

The most common measures, methods, and economic indicators of economically evaluating the return on capital investment are

1. Rate of return or return on investment (ROI)
2. Payment period (PP)
3. Discounted cash-flow rate of return (DCFR) and present value index (PVI)
4. Net present value (NPV)

9.5.4.1 The Annual Rate of Return

It is defined by the equation:

$$\text{ROI} = \frac{\text{Annual profit}}{\text{Capital investment}} 100$$

For projects, where the cash flow extends over a number of years, the average rate of return is calculated using an average value for the profit, by dividing the sum of the annual profits by the useful lifetime:

$$\text{ROI} = \left[\frac{\sum_{1=y}^{n} \text{Annual profits/n}}{\text{(capital investment)}}\right] 100$$

9.5.4.2 Payout Period, Payback Time, or Cash Recovery Period

Payout period is defined as the time required for the recovery of the depreciable capital investment in the form of cash flow to the project. Cash flow would imply the total income minus all costs except depreciation.

Mathematically, this is given by Equation 6.2, where the interest charge on capital investment is neglected:

$$\text{Payout period (years) (PP)} = \frac{\text{Depreciable capital investment}}{\text{Average annual cash flow}}$$

9.5.4.3 Discounted Cash Flow Rate of Return and Present Value Index

If we have an oil asset (oil well, surface treatment facilities, a refining unit, etc.) with an initial capital investment "P," generating annual cash flow over a lifetime n, then the DCFR is defined as the rate of return or interest rate that can be applied to yearly cash flow so that the sum of their present value equals P.

From the computational point of view, DCFR cannot be expressed by an equation or formula, similar to the previous methods. A three-step procedure involving trial and error is required to solve such problems, as illustrated in the solved example in Example 9.3.

9.5.4.4 Net Present Value

The DCFR method is based on finding the interest rate that satisfied the conditions implied by the method. Here, we provide a value for i that is an acceptable rate of return on the investment and then calculate the discounted value (present value) of the cash flow using this i. The net present value is then given by

$$\text{NPV} = (\text{present value of cash flow discounted at a given i}) - \text{capital investment}$$

END-OF-CHAPTER SOLVED EXAMPLES

Example 9.1

In a gas plant, a stream of natural gas is processed in order to recover the valuable hydrocarbons C_2 and C_3/C_4 and to sell the bulk of methane (C_1) as a fuel gas, sales gas.

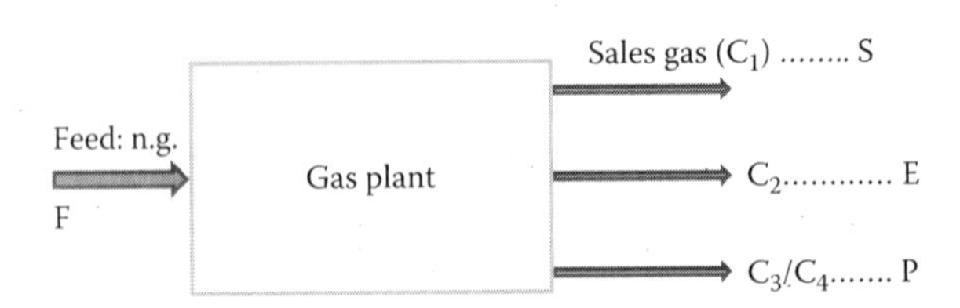

The feed rate is 2.5×10^6 g·mole/h. The feed gas composition is as follows:

Component	Mole%
C_1	83.0
C_2	12.0
C_3	4.0
n-C_4	1.0

Product specs are the following

- Sales gas contains negligible C_3/C_4.
- C_2 product contains 1.0 mole% C_1, 2.0 mole% C_3, and no C_4.
- C_3/C_4 product contains no C_1, 2 mole% C_2, and all of n-C_4.
  - Calculate the flow rates for each of the product streams, assuming that the sales gas is sold with 0.0 mole fraction of C_2.
  - Repeat for 0.01 and 0.05 mole fraction of C_2.

Solution

Mole balance will give the following

$$\text{For } C_1 \quad 0.83F = y_1S + 0.01E + 0 \tag{9.1.1}$$

$$\text{For } C_2 \quad 0.12F = y_2S + 0.97E + 0.02P \tag{9.1.2}$$

$$\text{For } C_3 \quad 0.04F = 0 + 0.02E + y_3P \tag{9.1.3}$$

$$\text{For } C_4 \quad 0.01F = 0 + 0 + y_4P \tag{9.1.4}$$

Given

$$y_1 + y_2 = 1.0 \tag{9.1.5}$$

$$0.02 + y_3 + y_4 = 1.0 \tag{9.1.6}$$

We have six equations and seven unknowns; F is given. Since the value of C_2 in the sales gas, y_1, is specified to be 0.0, the solution is obtained readily as follows:

Answer

$$S = 2.07 \times 10^6$$

$$E = 0.31 \times 10^6$$

$$P = 0.12 \times 10^6$$

Example 9.2

In the process design of a chemical plant, it is required to calculate the unknown mass flow rates of each outlet stream for the given separation system. The inlet mass flow rate (kg/h) and the mass fractions of each species in the inlet (stream 1), F, and each outlet are known.

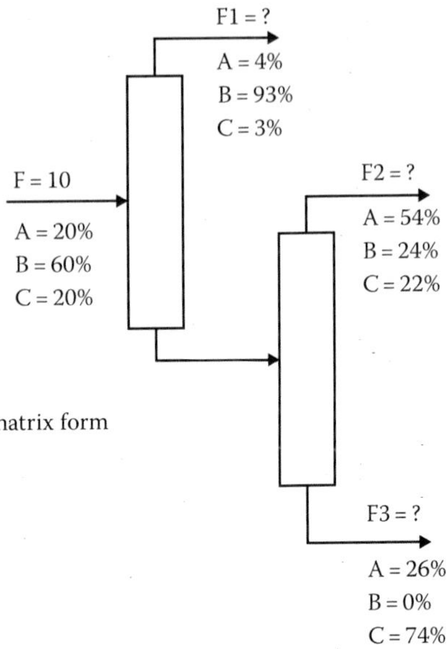

Solution

If we define the unknowns as
x1 = F1, x2 = F2, x3 = F3
and set up the mass balances for

1. The total mass flow rate
   x1 + x2 + x3 = 10

2. The mass balance on species 1
   0.04x1 + 0.54x2 + 0.26x3 = 0.2*10

3. The mass balance on species 2
   0.93x1 + 0.24x2 = 0.6*10

These three equations can be written in matrix form

F = 10: A = 20%, B = 60%, C = 20%
F1 = ?: A = 4%, B = 93%, C = 3%
F2 = ?: A = 54%, B = 24%, C = 22%
F3 = ?: A = 26%, B = 0%, C = 74%

$$\begin{bmatrix} 1 & 1 & 1 \\ 0.04 & 0.54 & 0.26 \\ 0.93 & 0.24 & 0 \end{bmatrix} \begin{bmatrix} x1 \\ x2 \\ x3 \end{bmatrix} = \begin{bmatrix} 10 \\ 2 \\ 6 \end{bmatrix}$$

To find the values of unknown flowrates write the code:

```
A = [1,1,1; .04, .54, .26; .93, .24,0] ;  B = [10; .2*10; .6*10];

X = A \ B;  F1 = X(1),  F2 = X(2),  F3 = X(3)
```

The results will be

```
F1 = 5.8238

F2 = 2.4330

F3 = 1.743
```

Example 9.3

The capital cost of a heat exchanger, P, is $10,000 with a lifetime of 10 years. If the money can be invested at 6% (annual interest rate), calculate

(a) The annual depreciation costs
(b) The annual capital recovery, if P is deposited in a bank at 6%
(c) Compare between the two values

Solution

(a) Calculation of $A_d = 10{,}000[(0.06)/\{(1.06)^{10} - 1\}] = \$759.0$
(b) Calculation of $A_r = P\{[i(1 + i)^n]/[(1 + i)^n]\} = 10{,}000(0.1359) = \$1{,}359.0$

(c) The difference between the two values is = 1359–759 = \$600; the interest on the capital, if invested in a sinking fund, gives 600(13.18) = \$7911

The A_{r}, on the other hand, if invested in sinking fund, should give

$$1359(13.18) = \$17,911,$$

which should break even with P, if deposited in a bank:

$$10,000(1+0.06)^{10} = \$17,911$$

Example 9.4

Assume an oil company is offered a lease of oil wells that would require a total capital investment of \$110,000 for equipment used for production. This capital includes \$10,000 working money, \$90,000 depreciable investment, and \$10,000 salvage value for a lifetime of 5 years.

Cash flow to project (after taxes) gained by selling the oil is as given next. Based on calculating the DCFR, a decision has to be made: should this project be accepted?

Year	Cash Flow ($\$10^3$)
0	−110
1	30
2	31
3	36
4	40
5	43

Solution

Our target is to set the following equity: by the end of 5 years, the future worth of the cash flow recovered from oil sales should break even with the future worth of the capital investment, had it been deposited for compound interest in a bank at an interest rate i. This amounts to say that

$$F_o = F_b$$

where

F_b = 110,000(1 + i)5 for banking

$F_o = \Sigma_{i=1}^{5} F_i$ for oil investment that represents the cash flow to the project, compounded on the basis of end-of-year income; hence

$$F_o = 30,000(1+i)^4 + 31,000(1+i)^3 + 36,000(1+i)^2$$
$$+ 40,000(1+i) + 43,000 + 20,000$$

Notice that the $20,000 represents the sum of WC and salvage value; both are released by the end of the fifth year. Setting up $F_B = F_o$, we have one equation involving i as the only unknown, which could be calculated by trial and error.

The value of i is found to be 0.207, that is, the DCFR = 20.7%.

Example 9.5

Calculate the NPV of the cash flow for the oil lease described in the example earlier, if the money is worth 15%.

Solution

At i = 0.15, the annual cash flow is discounted. The present value of the sum of the cash flows = $127,000.

$$NPV = 127,000 - 110,000$$
$$= \$17,000$$

That is, the oil lease can generate $17,000 (evaluated at today's dollar value) more than and above the totally recovered capital investment. Solution is illustrated in Table 9.1:

TABLE 9.1
DCFR: Results for Solved Example 9.4

		i = 5%		i = 20%		i = 25%		i = 20.7%	
Year (y)	Cash Flow	dy	Present Value ($)	dy	Present Value ($)	dy	Present Value ($)	dy	Present Value ($)
0	110,000								
1	30,000	0.8696	26,088	0.8333	24,999	0.8000	24,000	0.8290	24,870
2	31,000	0.7561	23,439	0.6944	21,526	0.6400	19,840	0.6870	21,297
3	36,000	0.6575	23,670	0.5787	20,833	0.5120	18,432	0.5700	20,520
4	40,000	0.5718	22,872	0.4823	19,292	0.4096	16,384	0.4720	18,880
5	43,000	0.4971	21,375	0.4019	17,282	0.3277	14,091	0.3910	16,813
	20,000								
	Total		117,444		103,932		92,747		102,380
	P.V.I		1.07		0.94		0.84		0.93
	N.P.V		$7,444.40		–$6,067.70		–$17,252.90		–$7,620.00

10 Case Studies

10.1 INTRODUCTION

This chapter is best introduced by recalling the definition that chemical engineering is the field that applies the core scientific disciplines of chemistry, physics, biology, and mathematics to transform raw materials or chemicals into more useful or valuable forms, invariably in processes that involve chemical change.

Case Studies in Chemical Engineering is one of a set of resources published by the chemical company ECI between 1971 and 1982. The booklets were the result of a series of workshops at which contributors from industry and schools collaborated.

The booklet is perhaps only of historical interest, but it does show how the principles of chemical engineering were applied to improve the existing processes and devise new ones in the days before the principles of green chemistry had been enumerated.

CACHE (Computer Aids for Chemical Engineering; CACHE Corporation, 2016) Process Design Case Studies represents an outstanding resource of many case studies. These are covered in 11 different volumes that span "Separation System for Recovery of Ethylene and Light Products from Naphtha Pyrolysis Gas Steam" in Volume 1, all the way to "Conceptual Design of an Aromatics Plant from Shale Gas" in Volume 2.

In Chapter 10, we attempt to introduce a number of case studies that touch on practical technical problems, some of which the author has personally experienced in his research work, both in academic institutions and in industry. Many of the principles and fundamentals presented in the preceding chapters of our book are introduced and illustrated as given in these case studies. Case studies 5 and 6 are taken from the open literature. The list of case studies presented includes the following:

- Case Study 1: Alternatives to Storage of Ammonia
- Case Study 2: Flash Dewatering of Raw Sewage Effluents
- Case Study 3: Magnesium Extraction from Seawater
- Case Study 4: Chemical Desalting of Brines: An Avenue toward Global Sustainability
- Case Study 5: Reactive Distillation
- Case Study 6: Applications of Differential Equations

10.2 CASE STUDY 1: ALTERNATIVES TO STORAGE OF AMMONIA

10.2.1 Introduction

In a process proposed for desert development by Abdel-Aal, ammonia is to be synthesized by using *solar hydrogen* produced by water electrolysis, as shown in Figure 10.1.

For this process, storage tank facilities are needed to store ammonia. It is estimated to store the amount of 60,000 tons of ammonia yearly.

10.2.2 The Problem

Outline the possible alternatives to store the ammonia; propose the one you recommend most. Carry out the calculations you deem necessary to support your recommendations.

10.2.3 Approach

The possible alternatives are

- Liquid at ambient temperature under its vapor pressure
- Liquid at atmospheric pressure refrigerated to its boiling point
- A solution in some solvent, say water, which lowers the vapor pressure and allows less elaborate containment

Step 1: Since ammonia is a vapor under normal conditions, one might store it in large gasholders. This idea is ruled out because of the excessive volume of ammonia at standard T and P, as shown next.

$$V = (60{,}000 \text{ tons})(2000 \text{ lb/ton})(1/17)\text{lb-mole/lb})(370)\ \text{SCF/lb-mole} = 3\times 10^3\ \text{ft}^3$$

Comment: A tank 100 ft high and a *mile* in diameter is not sufficient!

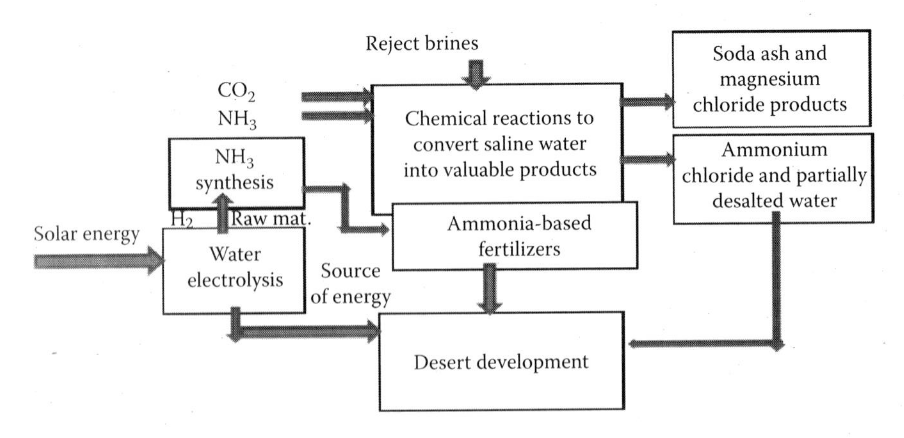

FIGURE 10.1 From solar hydrogen to desert development using ammonia.

Step 2: Let us calculate the vapor pressure of liquid ammonia at low temperatures.

Vapor Pressure (atm)	Temperature (°F)
1	−28
5	41
10	78

Let us assume that ammonia can be stored as a liquid at its normal boiling point. Then, we have to take into consideration the heat losses from ammonia tank to the atmospheric conditions.

Step 3: The primary resistance to heat transfer would be offered by the tank–air interface, with h equal to about 10 Btu/h · ft² · °F and k, of an insulation, is assumed to be equal to 0.15 Btu/h · ft · °F.

The surface area of the tank exposed to the atmosphere is

$$A = (5\pi/4)(4V/\pi)^{2/3}$$

where

$$V = 60{,}000 \times 2{,}000 \times 1/50\ (\text{ft}^3/\text{lb}) = 2.4 \times 10^6\ \text{ft}^3$$

Therefore,

$A = 8.5 \times 10^2\ \text{ft}^2$

The heat transfer rate is estimated as a function of the insulation thickness X (inches):

$$Q = UA\ \Delta T = [1/10 + X/0.15]^{-1}(8.5 \times 10^6)\ \text{Btu/h}$$

Calculations are done as follows:

X, Insulation Thickness (in.)	Q, Transferred (Btu/lb)
0.0 (no insulation)	8.5×10^7
6.0	2.1×10^5
12.0	1.1×10^5

10.2.4 Conclusions

It is observed that the heat transfer Q to the ammonia storage tank would lead to excessive loss of ammonia. Even with a foot of insulation, about 10^5 Btu/h will be transferred into the tank. This will result in the boiling of about 200 lb of ammonia per h (1 Btu vaporizes 590 lb ammonia).

It is therefore suggested to use 6 in. insulation that reduces the heat loss to

$$Q = 2.1 \times 10^5\ \text{Btu/h}.$$

Safety precautions have to be applied and start-up purge plan has to be considered to eliminate hazards of explosion involved during filling. In addition, the storage capacity of 60,000 tons could be divided into two units: one for loading and the other for unloading.

10.2.5 Discussions

1. Consider other possibility to store ammonia in a solid compound, from which ammonia is released as demanded; is it a feasible proposal?
   Could large underground cavern be considered as another option?
2. An optimization model could be carried out to calculate the optimum insulation thickness for the storage tank. This could lead to a substantial savings when choosing a small insulation thickness.

10.3 CASE STUDY 2: FLASH DEWATERING OF RAW SEWAGE EFFLUENTS

10.3.1 Introduction

Sewage treatment is the process of removing contaminants from wastewater, primarily from household sewage such as toilet flush, bath tubs, and washing machines. As shown in Figure 10.2, the treatment process normally involves many steps. Most important is the primary treatment, where the bulk of sludge is separated from the water.

10.3.2 The Problem

The Janicki Bioenergy Omni Processor (2) started as a proof-of-concept project, funded by the Bill and Melinda Gates Foundation in 2013. It was originally designed to take in sewer sludge and primarily output electrical power. During the development, it became clear that making clean drinking water made the processor more economically viable, and a water treatment system was added.

It is known that sewage effluents contain more than 90% water. Basically, the dewatering process is simply a kind of evaporation.

As a chemical engineer, make use of the fundamentals you learned in unit operations to propose a scheme to carry out this treatment.

10.3.3 Proposed Scheme

The scheme proposed is illustrated in Figure 10.3 and a schematic outline is given in Figure 10.4 as well. It utilizes the concept of multistage flash evaporation (MSFE) commonly used in the desalination of seawater. Before this evaporation step is carried out, the water content of the sewage effluents could be reduced to half its initial volume by the application of the primary treatment as illustrated in Figure 10.2. Water collected at this stage could be utilized for irrigation.

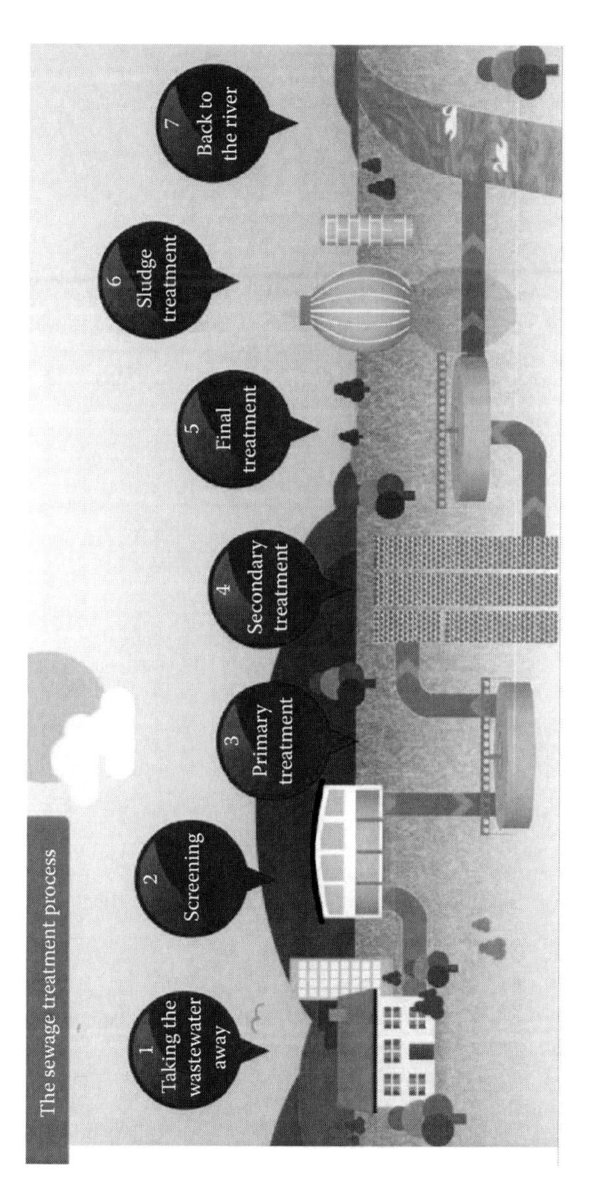

FIGURE 10.2 Sewage treatment process. (After Thames Water, *Facts and Figures*, http://www.thameswater.co.uk/about-us/4625.htm, retrieved on May 31, 2014.)

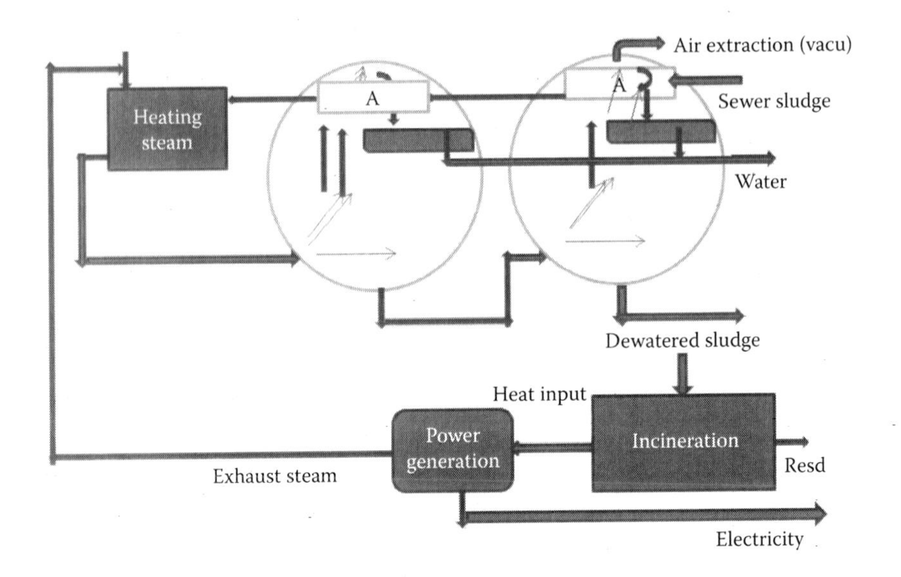

FIGURE 10.3 Proposed scheme for treating sewage sludge to produce water and electricity.

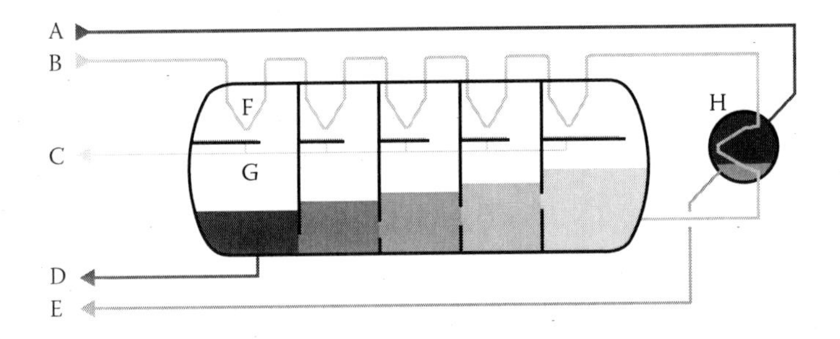

FIGURE 10.4 Schematic of a *once-through* multistage flash evaporator. *Notes:* A, steam in; B, sewage feed in; C, condensed water out; D, dewatered sludge out; E, condensate out; F, heat exchanger; G, condensation collection of water; H, sewage steam water heater (boiler). (Modified after Winter, T., Pannell, D.J., and McCann, L. *The Economics of Desalination and Its Potential Application in Australia*, SEA Working Paper 01/02, Agricultural and Resource Economics, Perth: University of Western Australia, 2001. http://www.general.uwa.edu.au/u/dpannell/dpap0102.htm, retrieved on November 13, 2006.)

10.3.4 Process Description

Sewage waste effluent, say the feed, is pumped into the heat exchanger (dryer) located in the top of the flashing chamber (stage 1), as shown in Figure 10.3. This way it gains heat from the uprising hot water vapor, which condenses as a product. The feed keeps moving to the next stage, which is then subjected for further heating inside the steam heater. As it enters stage 2, its temperature is above the boiling point at the pressure maintained at this stage. Therefore, a fraction of the water boils (*flashes*) to steam, thereby reducing the temperature until equilibrium is reached. The resulting steam is a little hotter than the feedwater in the heat exchanger. The steam cools and condenses against the heat exchanger tubes, thereby heating the feedwater as described earlier. By that time, the sewer sludge has been subjected to boiling, a process that removes all liquid, to be captured as water vapor upon flashing inside the chambers.

The water is collected through troughs located beneath the heat exchangers (A). Further treatment and ultrarefining steps are carried out for the produced water, making it suitable for drinking.

Dewatered sludge is directed to an incinerator, which produces high-pressure steam that fuels a generator. The generator creates electricity that is used as a source of power. There is even a little extra electricity left that can be transferred into the power grid. Exhaust steam, on the other hand, is used for heating purpose and for steam-jet ejector to create a vacuum.

Based on daily basis, the results of S200 Omni Processor are reported as follows:

1. Plant input
   a. A quantity of 10–12 tons of dry fuel
   b. It could be
      i. Any kind of biomass from sewage sludge
      ii. Animal by-products
      iii. Agricultural by-products
      iv. Garbage (glass- and metal-free)

      The fuel does not have to be dry as such. It can contain an appreciable amount of water: 12 tons of fuel can hold up to 70 tons of water.

      Arrangement can be made to have a feed input that contains a mixture of
      A. Sewage feed, having more than 90% water
      B. Sludge or other solid material, having 30%–90% water
2. Plant output

   Three items are produced:
   a. The power output is 300 kW.
   b. The water output has a total volume of 70,000 L (water produced will be clear and sterile, but with residual odor; ultrapurification equipment is required to make it odor-free).
   c. About 10%–20% of the dry material input will come out as ash.

10.3.5 Conclusions

The proposed scheme presented in our chapter consists of two stages. However, the number of flashing units (chambers) could multiple. This number is a function of the water content and the type of sludge. The advantage of applying vacuum in the first unit is obvious to reduce the temperature in the next units causing flashing of the entering feed. In addition, reduction of vacuum inside the chambers leads to lower operating pressure, and hence lower operating temperatures as seen in Figure 10.5.

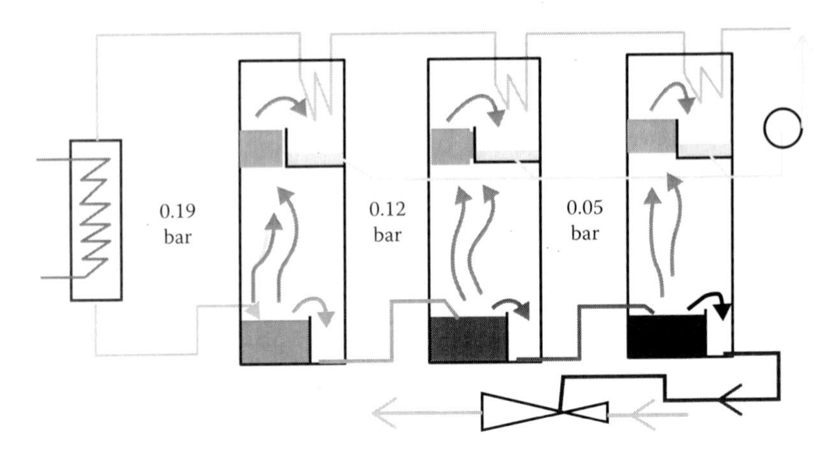

FIGURE 10.5 A module for multistage flashing operation.

Additional economic advantages are obtained when solar energy is considered in heating and boiling instead of steam.

10.3.6 Discussions

1. Derive the flash evaporation equation that applies for a single component using a heat balance to predict how much is vaporized from the feed to the evaporator. In this case, it will be assumed that the feed contains the solid material (A) suspended in a bulk of water, similar to seawater with soluble sodium chloride.
2. Elaborate on the incineration process. What are the basic design parameters to consider in building an incineration furnace?

10.4 CASE STUDY 3: MAGNESIUM EXTRACTION FROM SEAWATER

10.4.1 Introduction

The availability of magnesium metal from seawater plays a significant role in making such a strategic metal to compete with other used metals. It is only two-thirds as dense as aluminum. It is easily machined, cast, forged, and welded. It is used extensively in alloys, chiefly with aluminum and zinc and with manganese.

It is worth mentioning that magnesium is ranked as the eighth most abundant element in the Earth's crust. One cubic kilometer of seawater contains a minimum of one million tons of magnesium, which makes the sea a *storehouse* of about 1.7×10^{24} tons. Magnesium (Mg) is found in seawater as ions of magnesium. It is typically extracted from seawater on an industrial scale in a process known as the Dow process by precipitating it as magnesium hydroxide and then converting it to magnesium chloride using hydrochloric acid.

10.4.2 Proposed Method

In the proposed method, known as preferential salt separation (PSS), magnesium chloride is directly obtained from seawater and is used as a feedstock to manufacture magnesium. The whole scheme is illustrated as shown in Figure 10.6.

The extraction of magnesium chloride from seawater is done first, using solar energy to vaporize a dynamic stream of preconcentrated seawater flowing along an inclined preferential salt separator (PSS). Magnesium chloride salts soluble in seawater will separate as the very end product. Distilled water will be produced as a by-product in this phase. It represents a typical feed of water for hydrogen production by electrolysis.

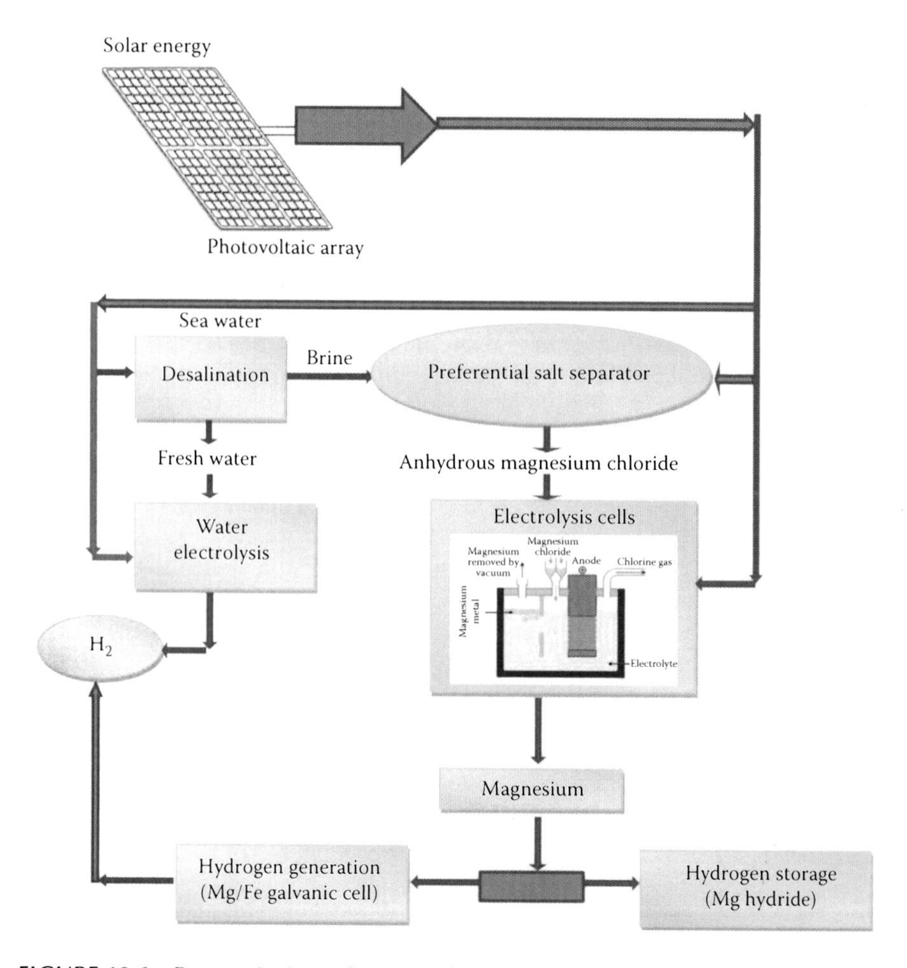

FIGURE 10.6 Proposed scheme for magnesium production from seawater.

10.4.3 Production of Magnesium Metal from Magnesium Chloride

Anhydrous magnesium chloride is electrolyzed next, using the energy generated by solar power in order to produce magnesium metal. Once produced, magnesium represents a reliable source of stored energy that could be exported by air, sea, or other means of transportation to remote locations for the production of hydrogen. Another option is to use magnesium to store hydrogen in the form of magnesium hydride. Recent advances in system integration enable a sustainable and stable power supply from solar systems.

10.4.4 Discussions

1. Draw a block diagram for the aforementioned process, indicating the type of operation in each step: physical (unit operation) or chemical conversion (unit process).
2. Identify the chemical reactions that take place.
3. Compare between the proposed method of extracting Mg chloride from seawater (PSS) and the one in current practice by Dow Chemical Company.

10.5 CASE STUDY 4: CHEMICAL DESALTING OF BRINES

10.5.1 Introduction

Innovations in chemistry have a great effect on separation and processing in chemical industries. As a matter of fact, many important chemical revolutions center about the discovery of new reaction paths, such as the one we are considering in this paper.

Solvay process is taken into account, which is referred to as the ammonia–soda process. It is the major industrial process for the production of *soda ash* (sodium carbonate). The ammonia–soda process was developed into its modern form by Ernest Solvay during the 1860s. The ingredients for this process are readily available and inexpensive: salt brine (from inland sources or from the sea) and limestone (from mines). A similar approach was followed by Abdel-Aal [1,2], but with different objectives. The target is to desalt high-saline water resources using ammonia and carbon dioxide to produce partially desalted water along with soda ash and other chemicals.

Along this line, it could be stated that while the soda ash is the main product in the Solvay process, it is a by-product in the proposed chemical desalting process. The main product is partially desalted water that contains ammonium chloride, which may be called *fertile* water, which will then be used for agricultural purposes.

10.5.2 Methodology

In its general term, desalting refers to a water treatment process that removes salt from saline water resources. It is simply represented as follows:

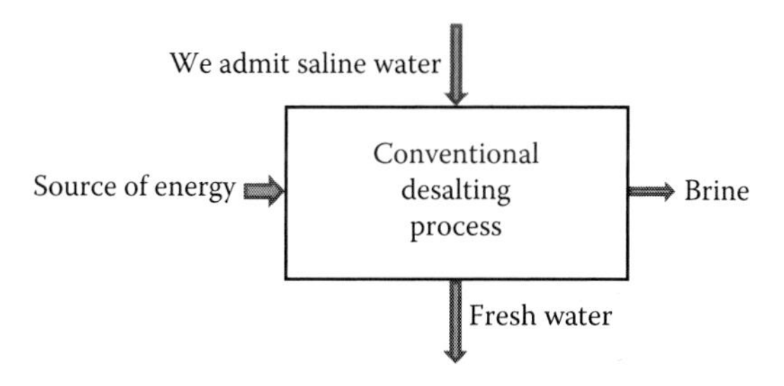

The proposed chemical desalting process, on the other hand, when compared to the conventional desalting process, differs on the following grounds:

1. Two chemicals, CO_2 and NH_3, are used in lieu of energy.
2. The feed is saturated brine or natural salt rocks.
3. Two main products are obtained:
   a. Partially–desalinated water that contains ammonium chloride (NH_4Cl) to produce *fertile water*
   b. Sodium carbonate (Na_2CO_3), which is known as soda ash

The process is represented schematically by the following scheme:

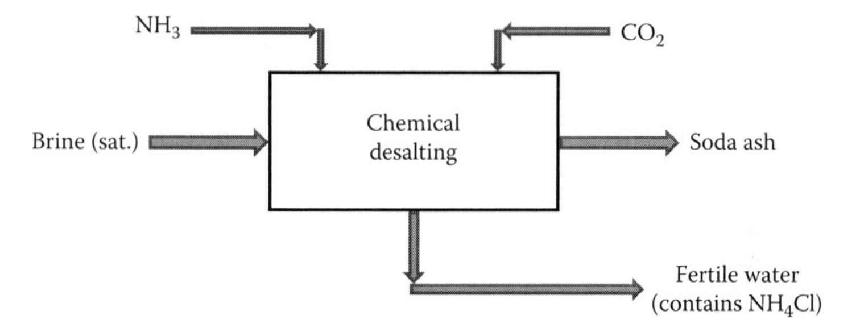

It could be safely stated that the chemical desalting process complements seawater desalination.

10.5.3 Main Reactions

The basic reactions involved could be visualized to take place as follows:

- *Primary reaction*: The reaction between CO_2 and NH_3 is described as follows:

$$CO_2 + NH_3 \rightarrow NH_2COOH \left[\text{carbamic acid}\right] \tag{10.5.1}$$

$$NH_3 + NH_2COOH \rightarrow NH_4^+ + NH_2COO^- \text{ [carbamate]} \tag{10.5.2}$$

$$\text{The net reaction is } CO_2 + 2NH_3 \rightarrow NH_2COO^- + NH_4^+ \tag{10.5.3}$$

- *Secondary reaction*: In the bulk of the solution, the carbamate hydrolyzes comparatively slowly to bicarbonate:

$$NH_2COO^- + H_2O \rightarrow NH_3 + HOCOO^- \quad (10.5.4)$$

- *Product formation reaction*
  Now, in the presence of NaCl, the following instantaneous reaction takes place:

$$NH_4 + HCO_3^- + NaCl \rightarrow NaHCO_3 + NH_4Cl \quad (10.5.5)$$

This leads to the precipitation of sodium bicarbonate leaving ammonium chloride in a partially desalinated solution.

10.5.3.1 Process Synthesis with Modifications

Sea oceans are a virtually inexhaustible source of magnesium. About 1 lb of magnesium is recovered from each 100 gal of seawater. Adding ammonia to our system will trigger the precipitation of magnesium as magnesium hydroxide $Mg(OH)_2$, which is separated as an intermediate product, as shown in Figure 10.7. This is a turning point in our process synthesis that will lead to the formation of NH_3 to be recycled as given by the following equation:

$$2NH_4Cl + Mg(OH)_2 \rightarrow Mg(Cl)_2 + 2NH_3 + 2H_2O \quad (10.5.6)$$

The separation of magnesium chloride as a product adds an economic value to the process.

10.5.3.2 Consumption–Production Analysis

The process involves the following reactions:

Reaction 1: $4NH_3 + 4H_2O \rightarrow 4NH_4OH$ (use 4 moles of NH_3)

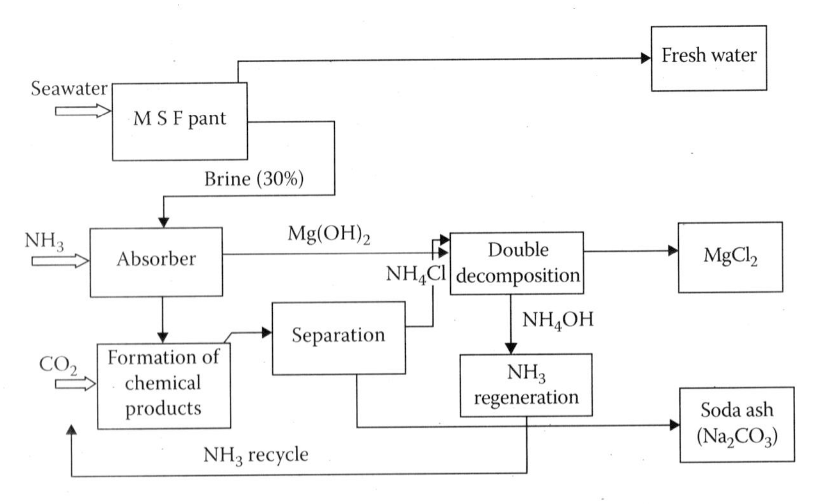

FIGURE 10.7 Block diagram for the proposed process.

Reaction 2: $2NH_4OH + Mg(Cl)_2 \rightarrow Mg(OH)_2 + 2NH_4Cl$

Reaction 3:

$2NH_4OH + 2CO_2 + 2NaCl \rightarrow 2NaHCO_3 + 2NH_4Cl$ (use 2 moles of NH_4OH)

Reaction 4: $2NaHCO_3 \rightarrow Na_2CO_3 + CO_2 + H_2O$

The consumption–production analysis is figured out as indicated in Table 10.1:

For process synthesis, the mass flow rates of the raw materials in and the products out could be readily calculated using any convenient basis for the flow rate input of saline brine containing 25%–30% sodium chloride as a saturated solution.

It should be noted that the number of moles of H_2O shown earlier are provided by the saline water upon admitting the ammonia gas into it. For CO_2 gas, it is recommended to use 2 moles as given in Equation 10.5.3. Partially–desalted water W, and is shown by * in Table 10.1, is obtained as a product, after separating the soda ash and magnesium chloride. If reaction given by Equation 10.5.6 applies to generate NH_3 for recycling, $MgCl_2$ is produced in this case.

10.5.3.3 Comparison between the Solvay Process and the Proposed Process

This comparison is done along the following lines:

Parameters	Solvay	Proposed
RM	Salt brine (rock salt)	Salt brine/desalination brines
	Limestone[+]	CO_2*
	Metallurgical coke	NH_3 (recycle)

$$NH_3\,(\text{recycle})$$

$$NH_3\,(\text{regeneration})\,2NH_4Cl + Ca(OH)_2 \rightarrow 2NH_3 + CaCl_2 + 2H_2O$$

$$2NH_4Cl + Mg(OH)_2 \rightarrow 2NH_3 + MgCl_2 + 2H_2O$$

Products	Soda Ash	Soda Ash
		$MgCl_2$
		Partially desalted water, W

10.5.3.3.1 Important Notes

- For Solvay process, metallurgical coke burns limestone, $CaCO_3$, to give $CaO + CO_2$ quick lime; CaO is slacked by water: $CaO + H_2O \rightarrow Ca(OH)_2$.
- For the proposed process, the source of CO_2 is the combustion of fossil fuels in power generation and water desalination plants.

TABLE 10.1
Consumption–Production Analysis for the Chemical Desalting Process

	Reactants					Products						
Reactions	NH_3	H_2O	CO_2	$MgCl_2$	NaCl	NH_4OH	$Mg(OH)_2$	$NaHCO_3$	Na_2CO_3	NH_4Cl	CO_2	W
No. 1	−4	−4				+4						
No. 2				−1		−2	+1		+2			
No. 3			−2		−2	−2		+2		+2		
No. 4								−2		+2	+1	
Net	−4	−4	−1	−1	−2	0	+1	0	+2	+4		*

10.5.4 Conclusion

The proposed process offers a scheme that provides three products as compared to one product only by the Solvay process. Magnesium chloride is an important product for the manufacture of magnesium metal. As a matter of fact, one can claim that the proposed process could compete with the well-known Dow process for the extraction of magnesium chloride from seawater.

The process synthesis suggested in this paper focuses on the use of ammonia as a recycle reagent. Regeneration of ammonia is accomplished in the absence of $Ca(OH)_2$ used in the Solvay process as indicated in the earlier comparison.

The production of fertile water (partially–desalted water) containing NH_4Cl could be a feasible choice; NH_4Cl could be sold as a solution to be used in fertilizers for rice crops.

10.5.5 Discussion and Problem Formulation

1. Check the values presented in Table 10.1 for the consumption–production analysis for the chemical desalting process.
2. Do a material balance for the process to calculate the quantities of the feed raw materials and the products obtained by processing a feed of 20×10^6 gallon per day (GPD) of seawater to be flashed, first in the MSF plant to produce 15×10^6 GPD of brine, with a salt concentration of about 7%.

   Waste–heat energy is used for salt concentration to raise the salt content in brine to the saturation point (about 30% concentration). This is the feed to the proposed process.

   Hint: The material balance on this concentrator is carried out as follows:

$$\text{Total balance } F = D + B$$

The component mass balance on the salt can be obtained as follows:

$$15 \times 10^6 (0.07) = B(0.3)$$

Therefore, $B = 3.5 \times 10^6$ GPD is the rate of concentrated brine to be processed next for the chemical conversion step, and a distillate rate of the freshwater is obtained by using the equation $D = 11.5 \times 10^6$ GPD.

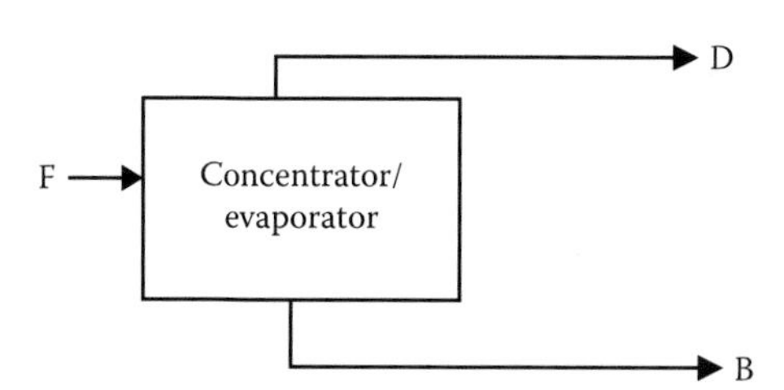

The main objective in this case study is to figure out how much chemical products we produce for a given feed input of brine. Also, it is important to calculate how much ammonia and carbon dioxide we consume in the process.

3. Comment on the proposed process as compared to the traditional water desalination.

10.6 CASE STUDY 5: APPLICATIONS OF DIFFERENTIAL EQUATIONS

10.6.1 Introduction

Many problems in engineering and science can be formulated in terms of differential equations. A differential equation is an equation involving a relation between an unknown function and one or more of its derivatives. Equations involving derivatives of only one independent variable are called ordinary differential equations. Since there are relatively few differential equations arising from practical problems for which analytical solutions are known, one must resort to numerical methods.

10.6.2 Problem Statement

Soap is prepared through a reaction known as saponification. In saponification, tallow (fats from animals such as cattle) or vegetable fat (e.g., coconut) is reacted with potassium or sodium hydroxide to produce glycerol and fatty acid salt known as *soap*. The soap is separated from the glycerol through precipitation by the addition of sodium chloride. Water layer on top of the mixture that contains dissolved sodium chloride is drawn off the mixture as a waste. This method of soap making is still being practiced in many villages in developing countries where the price of mass-produced soap maybe too expensive for the average villager.

10.6.3 Proposed Approach

Two chemical engineering students used their knowledge of saponification acquired in their organic chemistry class to organize and produce *homemade* soap. The local ordinance requires that the minimum concentration level for sodium chloride waste in any liquid that is discharged into the environment must not exceed 11.00 g/L. Sodium chloride laden liquid water is the major waste of the process. The company has only one 15 L tank for waste storage. On filling the waste tank, the tank contained 15 L of water and 750 g of sodium chloride. To continue production and meet the local ordinance, it is desired to pump in freshwater into the tank at the rate of 2.0 L/min, while waste saltwater containing 25 g of salt/L is added at the rate of 1.5 L/min. To keep the solution level at 15, 3.5 L/min of the waste is discharged.

A sketch in Figure 10.8 is given, where A represents the waste stream from the process, B is the freshwater stream, and C is the discharge stream to the

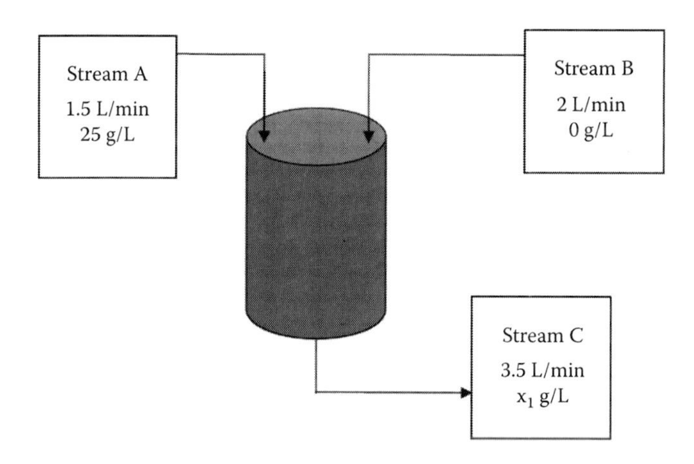

FIGURE 10.8 Flow diagram for a saponification process.

environment. Here, it is assumed that as the two streams, A and B, enter into the tank, instantaneously the chloride concentration in the tank changes to the exit concentration, x_1.

A material balance on sodium chloride is carried out for the tank system as follows:

$$\text{Rate of change in salt concentration with time} = \text{Input} - \text{Output} \tag{10.6.1}$$

$$\frac{dx_1}{dt} = (25\ \text{g/L})(1.5\ \text{L/min}) + (0\ \text{g/L})(2\ \text{L/min}) - (x_1\ \text{g/L})(3.5\ \text{L/min}) \tag{10.6.2}$$

$$\frac{dx_1}{dt} + 3.5x_1 = 37.5 \tag{10.6.3}$$

Divide both sides by 3.5

$$\frac{1}{3.5}\frac{dx_1}{dt} + x_1 = 0.286\frac{dx_1}{dt} + x_1 = 10.7 \tag{10.6.4}$$

By using an integrating factor (IF) for Equation 10.6.4, solution is found to be

$$X_1 = 10.7 + C\ e^{-t/0.286}$$

For the initial boundary conditions, at t = 0, the salt concentration in the tank was given as 750 g/15 L (50 g/L) or at t = 0, $x_1(0) = 50$ g/L.

Therefore,

$$50 = 10.7 + C, \quad \text{or} \quad C = 50 - 10.7 = 39.3$$

The final solution is given by

$$X_l = 10.7 + 39.3e^{-t/0.286}$$

Checking:
As t increases, $e^{-t} \rightarrow 0$, *and* $X_l \rightarrow$ a, value = 10.7
And at t = 0, $e^{-t} = 1$, *and* $x_l \rightarrow$ a, value = 50

10.6.4 Discussions

- Using Equations 10.6.3 and 10.6.4, numerically determine how the concentration of the discharged salt changes with time.
- Plot a graph for the relationship that represents your solution.
- How long did it take to achieve the minimum required local specifications? At steady state, what is the concentration of salt being discharged from this local soap factory? (Kaw and Kalu, 2011)

10.7 CASE STUDY 6: REACTIVE DISTILLATION

10.7.1 Introduction

Reactive distillation (RD) is a process in which a catalytic chemical reaction and distillation (fractionation of reactants and products) occur simultaneously in one single apparatus. Separation of the product from the reaction mixture does not need a separate *distillation* step, which saves energy (for heating) and materials. Reactive distillation or catalytic distillation combines reaction and distillation into a single column.

This technique is especially useful for equilibrium-limited reactions such as esterification and ester hydrolysis reactions. Conversion can be increased far beyond what is expected by the equilibrium due to the continuous removal of reaction products from the reactive zone. This helps reduce capital and investment costs and may be important for sustainable development due to a lower consumption of resources.

10.7.2 Objective

In comparing the traditional versus the reactive distillation methods, such as in the production of methyl acetate, for example (Figure 10.9), the latter methods have the following advantages:

- Improve selectivity
- Reduce by-products
- Prevent pollution
- Reduce energy use
- Handle difficult separations
- Avoid separating reactants

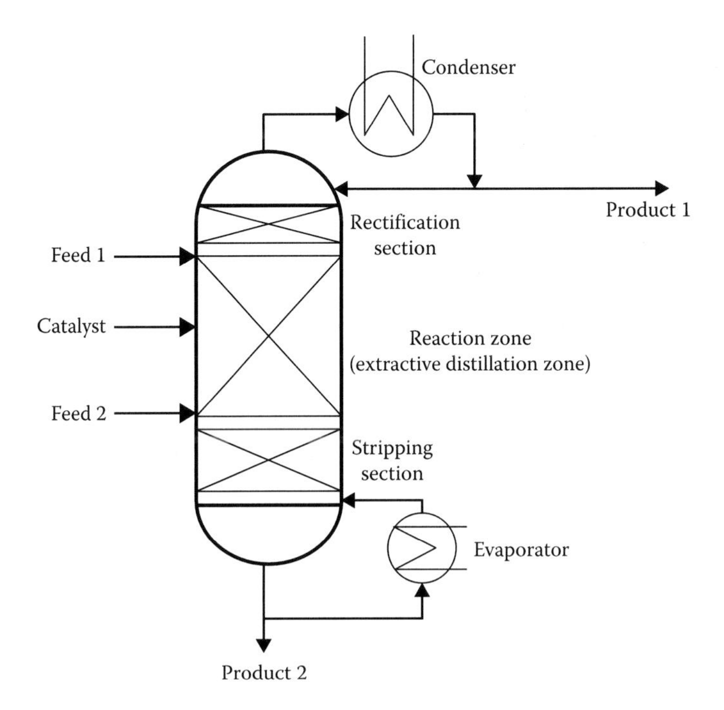

FIGURE 10.9 Example of production of methyl acetate.

- Eliminate/reduce solvents
- Enhance overall rates
- *Beat* low equilibrium constants

10.7.3 Discussion

The introduction of a separation process side by side in the reaction zone, or vice versa, could lead to many difficulties in reactive distillation process. Give examples of some of these problems. On the other hand, some of the merits that apply in the case of reactive distillation could be applied to other aspects in many unit operations. Give examples of similar cases, especially in surface oil operations that involve crude oil and gas field treatment and processing (see texts by the author in bibliography).

REFERENCES

CACHE Corporation, CACHE Process Design Case Studies, The CACHE Corporation, 2016. http://cache.org/super-store/cache-process-design-case-studies.

The Essential Chemical Industry (ECI), *Case Studies in Chemical Engineering*, Imperial Chemical Industries (ICI), London, U.K., 1970–1979.

Kaw, A. and Kalu, E.E., *Numerical Methods with Applications: Abridged*, 2nd edn. Tampa, FL: University of South Florida, March 2011.

Appendix A: Application Problems

A.1 SECTION I (CHAPTERS 1 AND 2)

This set of problems is aimed to go hand in hand with the materials covered in Chapters 1 and 2. The solution of some simple problems could be done either manually or using MATLAB® or Excel. It involves basic conversion of units and other fundamental calculations.

1. The heat capacity of sulfuric acid is measured in cal/(g · mol)(°C). It is given by

$$\text{Heat capacity} = 33.25 + 3.727 \times 10^{-2}\ \text{T},$$

where T is in °C

It is required to have the heat capacity expression in terms of °R.

2. A tank contains CO_2 is used to supply a system with CO_2. The pressure gauge reads 50 psi. Determine the absolute pressure in the tank, in psi, if the barometer reads 28.0 in. · Hg.
3. Calculate the gauge pressure in $lb_f/in.^2$ at a depth of 4.0 miles below the sea surface. Assume the water temperature is 60°F and the specific gravity is 1.042.
4. Calculate the diameter of a pipeline handling 12,000 bbl of oil/h assuming the velocity of flow in the pipe line is 5 ft/s (1 ft^3 = 0.413 barrel).
5. Polonium has a half-life of 140 days. Starting with 50 g today, how much is left after 200 days?

$$\text{Amount left} = \text{Initial amount}(0.5)^{\text{half-life time}}$$

6. Find the volume of a tubular reactor, with a residence time of the reaction being 60 s. The input feed flow rate to the reactor is 400 ft^3/min.
7. What is the weight of air in a room having a volume of 2000 ft^3 at T = 80°F and P = 750 mm Hg.
8. In one of your experiments in unit operations laboratory, a reading of 400 cm^3/min was measured. How do you covert it to $in.^3$/h?
9. In experiment in problem 8, the pressure gauge reads 12 psi and the atmospheric pressure is 14 psi. What is the absolute pressure in this experiment?

10. What will be the concentration of O_2 dissolved in water at 298 K when the solution is in equilibrium with air at 1 atm total pressure?
Hint: Apply Henry's law, $p_A = Hx_A$,

where
H is the Henry's constant = 4.38×10^4 atm/mol fraction
p_A is the partial pressure of O_2 in air, which is taken as 0.21 atm.

11. Find the density of methane gas at T = 70°F and P = 760 mm Hg, by using the gas equation.
12. An orifice meter is used to measure flow rates in a pipeline. Given the following relationship,

$$u = C\,[\Delta P/\ell]^{0.5},$$

where
u is the fluid velocity
ΔP is the pressure drop
ℓ is the density
C is a constant

What are the dimensions of C in the cgs system of units?
13. Crude oil when produced may contain water. For the separation of waterdrops from oil, it was found that V = f (d, T), where V is the settling velocity of the waterdrop, d is the drop diameter (in.), and T is the operating temperature. In charge of this separator, you are asked to plot the following data to indicate the relationship between these variables (*hint*: T versus V). Notice that V increases (separation is better) with higher T and larger d.
Settling Velocity, V (ft/h)
Values of V at different T and for different d are as follows:

T (°F)	d = 0.1	d = 0.15	d = 0.2
110	5	16	28
130	8	18	30
160	10	20	35

14. Analyze the following equation for dimensional consistency: $P = g \times h$, where g is the gravitational acceleration and h is the height of the fluid.
15. Hydrogen is produced by water electrolysis: $2H_2O \rightarrow 2H_2 + O_2$.
Calculate how much hydrogen is produced from 800 kg of water. *Hint*: Use mole balance.

A.2 SECTION II (CHAPTERS 3 AND 4)

This set of problems may require using MATLAB or Excel to solve them as demonstrated before. It is up to the instructor to assign these problems chapterwise, as application exercises while covering lectures.

16. A chemical plant located at the base of a mountain (about 300 ft) has to dispose of its CO_2 by-product. An engineer proposed a unique method for absorbing the CO_2 in water; water is then lifted up the mountain. The CO_2 was then vented to the atmosphere and water is recirculated as shown in the diagram. Assume that CO_2 is to be absorbed in water at a CO_2 partial pressure, p_{CO_2}, of 10 psi above that required to lift the water up to the top of the mountain (300 ft). Calculate the amount of water required to dispose of 5000 ft^3 of CO_2 at STP.

    Hint: (a) Calculate p_{CO_2} to take care of the pressure head for uplifting. (b) Apply Henry's law to find the X_{CO_2} absorbed in water: $p = Hx$.

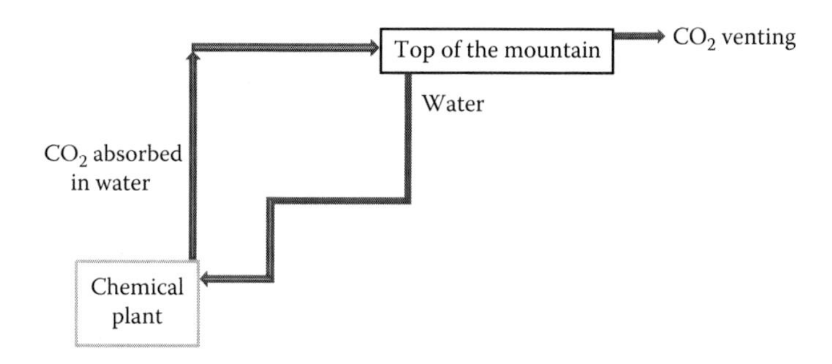

17. For the flash separation of a multicomponent mixture, show that $Y_i = Z_i/[(1/K_i) - (V/F)(1/K_i-1)]$.

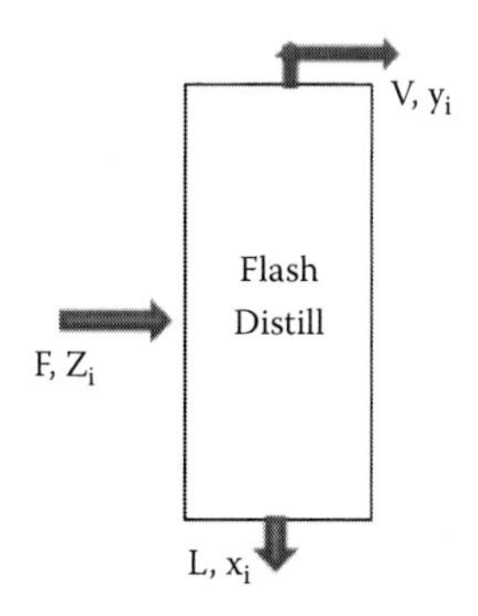

Hint: Make material balance (MB) and eliminate x_i using $y_i = K_i x_i$.

18. A stream of air is passing through a tube. Its flow rate, F, is unknown. In order to determine the value of F, another gas, CO_2, is introduced in the tube at the entrance point, as shown in the diagram at a rate = 1 ft^3/s. The composition of the effluent stream (air + CO_2) that exits the tube was measured and found to have 0.5% CO_2 by volume. Calculate F, using component MB on CO_2.

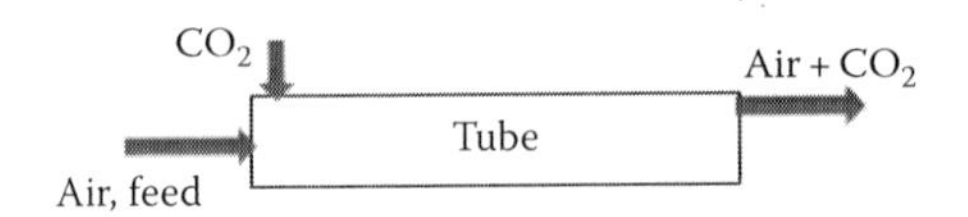

19. Seawater is desalinated using multistage flash distillation. Assuming the concentration of the seawater is C_0, the feed rate is 1000 gal/min, and the desalination process will produce 50% of the feed as top desalinated product and the balance as bottom product, figure out the concentration of this product, as a ratio of C_0.
20. A tank containing 8 ft^3 of a mixture of water and acetic acid has a valve at the bottom. If 40% of this mixture is acetic acid, how much of this mixture should be drained from the tank (through the valve) and to be replaced by pure acid in order to have the new concentration of the mixture be 60%?

    Hint: Make an MB on the acid. Let X be the amount of mixture to be drained and set up one algebraic equation in one unknown, X.
21. A continuous distillation column is to separate a mixture of two components A and B at atmospheric pressure. A feed is entering the column (bubble-point feed) at F moles/h, with X_{FA} as feed composition. Products are X_{DA} and X_{BA}. Reflux ratio is $R = L_0/D$, where L_0 enters the column at T°C.

    If h_F, h_D, H_v, and h_w are the enthalpy of feed, enthalpy of distillate, enthalpy of top vapor, and enthalpy of bottom product, respectively, show how you calculate Q_c and Q_R with the help of the diagram shown.

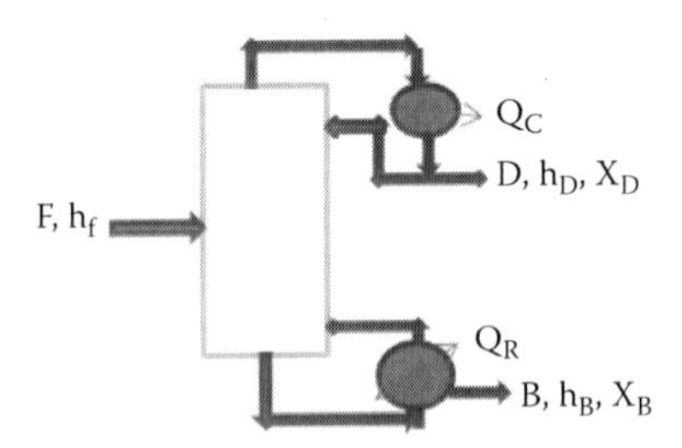

22. Applying the Newton's law of cooling, $dT/dt = -k(T-30)$, where the temperature of moving air is 30°C, T is the temperature of the substance at time t minutes, and k is a constant, find the time necessary to cool a substance from 100°C to 40°C, if the same substance cools from 100°C to 70°C in 15 min time.

23. Compute the bubble-point temperature (T_{bp}) at P = 2 atmosphere for the three-component mixture given next, using the Newton's method. Take the first assumed value for T_{bp} = 150°F.

Component	K_i	Composition (x_i)
1	$K_1 = 0.01T/P$	0.333
2	$K_2 = 0.02T/P$	0.333
3	$K_3 = 0.03T/p$	0.333

24. Estimate the average density of a water/ethanol mixture, to be prepared for distillation at different water compositions knowing that

$$\text{Water density} = 1000\ \text{kg/m}^3$$

$$\text{Ethanol density} = 780\ \text{kg/m}^3$$

Mixture density = Xwater (Water density) + Xethanol (Ethanol density)

25. A pilot plant distillation column is used to strip alcohol from a feed stream of ethyl alcohol and water, containing 1.5 mole% alcohol. The following conditions apply:

$$\text{Feed rate} = 16\ \text{lb} \cdot \text{mol/h},$$

where the distillate contains 87% alcohol and the bottom product contains 0.05 % alcohol. Calculate all stream flow rates.

26. At 1000°C and a total pressure of 30 atm, the equilibrium in the reaction

$$CO_2(g) + C(s) \rightarrow 2CO(g)$$

is such that 17 mole% of the gas is CO_2. What percent will be CO_2 if the total pressure is 20 atm instead of 30 atm?

27. Draw a computer block diagram for solving flash separation problem by trial and error.
28. Find the value of V/F for a feed consisting of 10, 20, 30, and 40 mole% of propane, butane, pentane, and hexane, respectively, to be flashed at 200°F and 100 psia. The corresponding values of K_i for the hydrocarbons are 4.5, 1.8, 0.75, and 0.37 (in the same order).
29. Calculate the temperature and composition of a liquid in equilibrium with a gas mixture containing 10.0 mole% benzene, 10.0 mole% toluene, and the balance nitrogen (which may be considered noncondensable) at 1 atm. Is the calculated temperature a bubble-point or dew-point temperature?
30. Under limited conditions, *volume balance* can be applied to liquids in conjunction with a *mass balance*. Non–polar mutually–soluble liquids in general are blended so that the volume is approximately equal to the sum of the

liquid volumes of the components. Find the yields of light and heavy cuts that can be produced from 100 gal of gas oil under the given conditions:

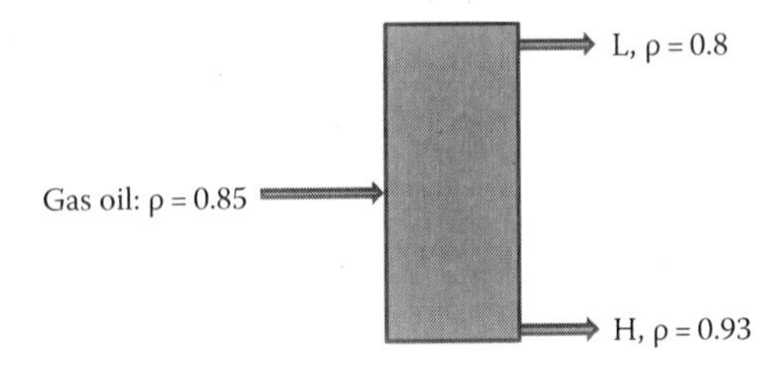

31. During a chemical reaction, substance A is converted into B at a rate proportional to the square of the amount of A:

$$A \rightarrow B.$$

If at time t = 0, 60 g of A is present, and then after 1 h (t = 1), only 10 g of A remains unconverted, find how much of A is present after 2 h?
Hint: Let C be the amount of A present at any time. Solve the differential equation: dc/dt = …

32. A mixture containing 45% benzene (B), 50% toluene (T), and 5% nitrobenzene by mass is fed to a distillation column. The overhead stream consists of 95% B and 4% T. The bottom product contains 90% T.
Assuming the feed rate is 2000 kg/h.
Calculate the overhead and bottom flow rates. Determine the mass flow rates of B and T in the bottom product.

33. Write a program to estimate the physical properties for water in the range of temperatures from 273 to 323 K.
Given
Density

$$\rho = 1200.92 - 1.0056T + 0.001084T^2$$

Conductivity

$$K = 0.34 + 9.278 * 10^{-4}T$$

34. Make a table of (x vs. y) values for the three components such as ethanol, water, and benzene, at T = 373.15 K. The vapor pressure of these three components is calculated by (*Note*: Take 11 points for each component from 0 to 1)

Ethanol $$P_e^o = \exp\left(\frac{18.5242 - 3578.91}{T - 50.5}\right)$$

Water $$P_w^o = \exp\left(\frac{18.3036 - 3816.44}{T - 46.13}\right)$$

Benzene $$P_b^o = \exp\left(\frac{15.9008 - 2788.51}{T - 52.36}\right)$$

where

$$K_i = P_i^o / P_t$$

$$P_t = 760$$

$$y_i = K_i \times x_i$$

35. An ammonia synthesis converter receiving a total feed (T moles) consists of H_2 and N_2 at a 3:1 ratio for the following reaction:

$$N_2 + 3H_2 \rightarrow 2NH_3$$

The total feed (T moles) is made up of fresh feed (F moles) + recycle stream (R moles). The gross product (G moles) is 12% NH_3 moles, which goes to a separator, removing all NH_3 and allowing recycled stream to return to the convertor. For F = 100, calculate G and R.

A.3 SECTION III (CHAPTERS 5 THROUGH 9)

Similar to the other two parts, a set of application problems is presented. This set comprises many types of problems frequently encountered by chemical engineering students.

36. In a distillation train, a liquid hydrocarbon containing 20 mole% ethane, 40 mole% propane, and 40 mole% butane is to be fractionated into essentially pure components as shown next.
    On the basis of F = 100 moles, what is the value of P (in moles) and the composition of stream A?

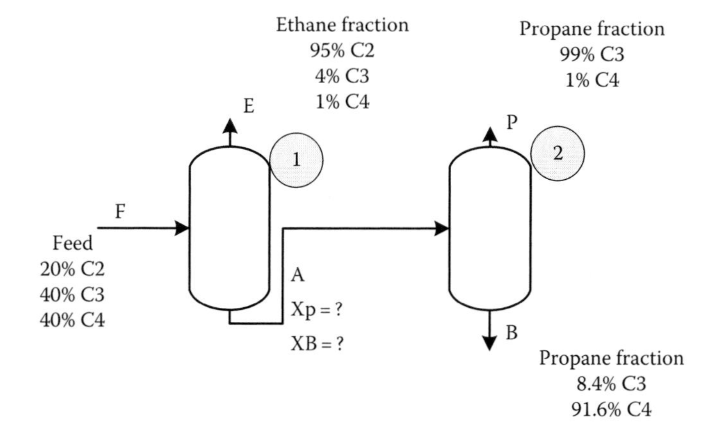

37. Equilibrium constants (k values) are needed for our distillation calculations, as was explained earlier. Write a program to calculate the vapor pressure for the following system and to calculate and tabulate its k values.

$$P_1 = \exp\left(\frac{14.2724 - 2945.47}{T + 224}\right)$$

$$P_2 = \exp\left(\frac{14.2043 - 2972.64}{T + 209}\right)$$

Temperature range = 75°C–125°C. Use 11 points for temperature

$$Pt = 101.3$$

$$K_i = \frac{P_i}{P_t}$$

38. Water at 320 K is pumped through a 40 mm ID pipe through a length of 150 m in a horizontal direction and up through a vertical height of 12 m at a rate of 2.16 m^3/h. In the pipe there are fittings equivalent to 260 pipe diameters. What power must be supplied to the pump if it is 60% efficient? Take the value of fanning friction factor as 0.008. Water viscosity is 0.65 cp and density = 1 gm/cc.
39. As a leader in your group, in charge of process design, you are asked to draw a simplified process flow diagram for the production of hydrochloric acid, HCl, as described next:

    It is manufactured by the reaction of NaCl and H_2SO_4. The raw materials are fed to a furnace where they are mixed and *indirectly* heated. The acid and salt react to give HCl gas, which then flows to heat exchangers to be cooled by water and then passes successively to gas-absorption towers. The gas enters the bottom of the first tower, leaves at the top, and flows into the bottom of the second tower. Water is fed to the top of the second tower, while the dilute HCl acid collected from the bottom of the same tower is supplied to the top of the first tower. Strong HCl acid is withdrawn at the base of the first tower and pumped to a storage tank. Waste gas is withdrawn by a blower at the top of the second tower.
40. Water at the rate of 150 lbm/min is heated from 100°F to 160°F using oil with a specific heat of 0.45. A countercurrent double-pipe heat exchanger is used. The oil enters at 230°F and leaves at 150°F. Assuming U_0 is 60 $Btu/h \cdot ft^2 \cdot °F$, calculate the heat exchanger area.
41. One face of a copper slab is maintained at 1000°F and the other face is at 200°F. How much heat is conducted through the slab per unit area?
    Given: The slab is 3 in. thick and $k = 215\ Btu/h \cdot ft \cdot °F$.

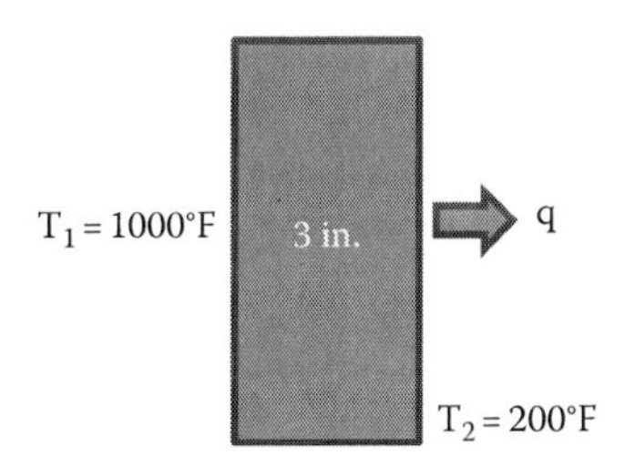

42. A solution of ethyl alcohol containing 8.6% alcohol by weight is fed at the rate of 5000 kg/h to a continuous fractionating column operating at atmospheric pressure. The distillate, which is the desired product, contains 95.4% alcohol by weight and the residue from the bottom of the column contains 0.1% alcohol by weight. Calculate the following:
    i. The mass flow rates of the distillate and residue in kg/h
    ii. The percentage loss of alcohol
43. A feed of 50 mole% hexane and 50 mole% octane is fed into a pipe still through a pressure reducing valve and then into a flash disengaging chamber. The vapor and liquid leaving the chamber are assumed to be in equilibrium. If the fraction of the feed converted to the vapor is 0.5, find the compositions of the top and bottom products. The following table gives the equilibrium data for this system:

Mole fractions of hexane in liquid x	1.00	0.69	0.40	0.192	0.045	0.00
Mole fractions of hexane in vapor y	1.00	0.932	0.78	0.538	0.1775	0.00

44. A continuous fractionating column is to be designed for separating 10,000 kg/h of liquid mixture containing 40 mole% methanol and 60 mole% water into an overhead product containing 97 mole% methanol and a bottom product having 98 mole% water. A mole reflux ratio of 3 is used. Calculate (i) moles of overhead product obtained per hour and (ii) number of ideal plates and location of the feed plate if the feed is at its bubble point.

    Equilibrium data are as follows:

x	0.1	0.2	0.3	0.4	0.5	0.6	0.7	0.8	0.9
y	0.417	0.579	0.669	0.729	0.78	0.825	0.871	0.915	0.959

where
x is the mole fraction of methanol in liquid
y is the mole fraction of methanol in vapor

Given: Feed rate = 10,000 kg/h, $z_F = 0.4$, $x_D = 0.97$, $X_W = 0.02$, R = 3

45. A saturated liquid mixture containing 60 mole% benzene and 40 mole% toluene is to be distilled continuously into a distillate product containing 90 mole% benzene and the bottom product containing 5 mole% benzene. The fractional distillation column will operate at approximately constant pressure of 1 atm. The reflux ratio is 2. How many theoretical plates must the column have if the feed is introduced into the eighth plate? Equilibrium data are as follows:

x	0	0.017	0.075	0.13	0.211	0.288	0.37	0.411	0.581	0.78	1
y	0	0.039	0.161	0.261	0.393	0.496	0.591	0.632	0.777	0.9	1

Given: $Z_F = 0.6$, $X_D = 0.9$, $X_W = 0.05$, $R = 2$, $X_D/(R + 1) = 0.9/(2 + 1) = 0.3$
Feed is saturated liquid at eighth plate.

46. Heat loss/gain takes place in a pipe carrying hotter/colder fluid than ambient temperature. Insulation reduces the heat loss to surroundings. Heat loss depends upon the number of factors like insulation thickness, ambient temperature, and wind speed. This shows how to calculate heat loss from an insulated pipe and a bare pipe to surroundings.

Problem statement: A carbon steel pipe, 3 in. diameter, is carrying hot oil at 180°C and insulated with 50 mm thick calcium silicate. Insulation is cladded with a sheet with surface emissivity of 0.9. Ambient temperature is 28°C and wind velocity is 3.5 m/s.

Calculate the surface temperature and heat loss from insulated and bare pipe.

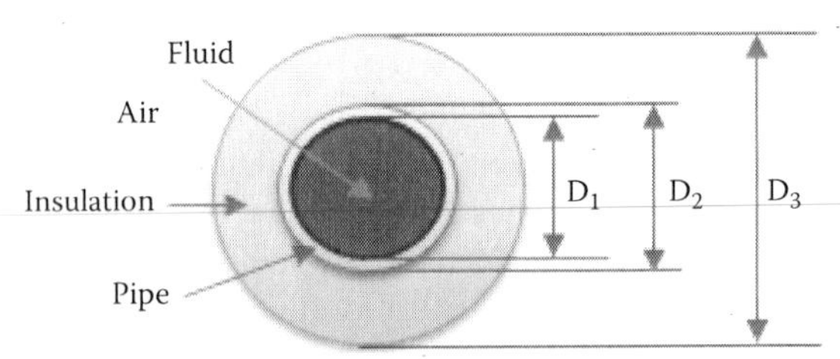

The overall heat transfer coefficient of an insulated pipe is defined as the follows:

$$U = \frac{1}{\dfrac{D}{D_1 \cdot h_{in}} + \dfrac{D_3 \cdot \ln\left(\dfrac{D_2}{D_1}\right)}{2 \cdot k_{PIPE}} + \dfrac{D_3 \cdot \ln\left(\dfrac{D_3}{D_2}\right)}{2 \cdot k_{INSULATION}} + \dfrac{1}{h_{AIR}}}$$

where

k_{PIPE} and $k_{INSULATION}$ are the thermal conductivities of the pipe and insulation

h_{in} is the heat transfer coefficient for fluid flowing in pipe

h_{AIR} is the heat transfer coefficient due to air flowing outside the pipe

The first two terms of the denominator in this equation are generally smaller compared to the remaining terms and can be neglected. For this example, the first term due to pipe fluid is ignored.

47. An evaporator is fed with 10,000 kg/h of a solution containing 1% solute by weight. It is to be concentrated to 1.5% solute by weight. The feed is at a temperature of 37°C. The water is evaporated by heating with steam available at a pressure of 1.34 atm absolute, corresponding to a temperature of 108.3°C. The operating pressure in the vapor space is 1 atm absolute. Boiling point elevation and other effects can be neglected. The condensate leaves at the condensing temperature. All the physical properties of the solution may be taken to be the same as that of water. What is the quantity of steam required per hour?

    Data

    Enthalpy of feed = 38.1 kcal/kg
    Enthalpy of solution inside the evaporator (at 100°C) = 98 kcal/kg
    Enthalpy of vapor at 100°C = 644 kcal/kg
    Latent heat of vaporization of steam = 540 kcal/kg

48. Methanol is produced by the following reaction:

$$CO + 2H_2 \rightarrow CH_3OH$$

    In the reactor, 15% of the CO is converted to methanol. Materials exit the reactor and flow to a separator to produce liquid methanol as a product. A recycle of unreacted gases is done. A purge stream is done for the impurities as shown in the diagram.

    Calculate the amount of feed, F_1, recycle, F_2, and purge, F_4. Assume 10,000 mol/h of methanol product, F_3.

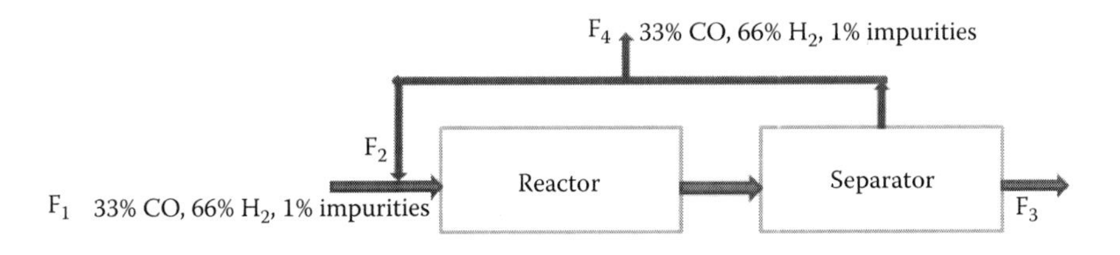

49. A triple-effect evaporator is concentrating a liquid. Temperature of the steam to the first effect is 227°F (about 5 psig). Vacuum on the last effect is 26 in. (125°F).

    Values of the overall heat transfer coefficients are 500, 400, and 200 for the first, second, and third effect, respectively. What are the approximate temperatures at which the liquid will boil in the first and second effect? *Hint*: Temperature drops will be approximately inversely proportional to the heat transfer coefficient.

50. Write a program to calculate the vapor pressure for the following system and to calculate and tabulate its k values.

$$P_1 = \exp\left(\frac{14.2724 - 2945.47}{T + 224}\right)$$

$$P_2 = \exp\left(\frac{14.2043 - 2972.64}{T + 209}\right)$$

Temperature range = 75°C–125°C. Use 11 points for temperature.

$$Pt = 101.3$$

$$K_i = \frac{P_i}{P_t}$$

A.4 ANSWERS TO PROBLEMS

1. Heat capacity = $23.06 + 2.07 \times 10^{-2}\ T_R$.
2. $P_{abs} = 13.75 + 50 = 63.75$ psi.
3. $P = 9533\ lb_f/in.^2$
4. D = 0.59 ft.
5. $N(t) = 50(1/2)^{200/140} = 50(1/2)^{1.4286} = 50 \times 0.3715 = 18.575$ g.
6. $V = (q)(\theta) = (400\ ft^3/min)(60\ s)(1/60)\ (s/min) = 400\ ft^3$.
7. Mass of air = $2000\ ft^3 \times 0.0735\ lb/ft^3 = 147$ lb.
8. $1608\ in.^3/h$.
9. 26 psi.
10. $x_A = (0.21)/4.38 \times 10^4 = 0.0479 \times 10^{-4}$.
11. $\rho = 0.0414\ lb/ft^3$.
12. Use cgs units.
13. Plot v versus D using T as a parameter.
14. Use basic units.
15. 88.9 kg hydrogen is produced from 800 kg water.
16. Quantity of water = 8730 lb = 3960 kg.
17. Use simple substitution.
18. F, feed rate = $11{,}940\ ft^3/min$.
19. Solve for $C_B = (1000/500)C_0 = 2C_0$.
20. $X = 8/3\ ft^3$.
21. Q_r (heat duty for reboiler) = $Dh_D + Bh_B + V_1H_v - D[h_D + (L_0/D)] - (F)h_f$.
22. t = 52 min.
23. Follow the solution of Example 4.1.
24. Solution is done through MATLAB.
25. $D = 0.267\ lb \cdot mole/h$
    $B = 15.733\ lb \cdot mole/h$.
26. Mole% $CO_2 = (2.55/20)100 = 12.75$.
27. Solution follows the Newton–Raphson procedure.
28. V/F is found = 0.1765.
29. Solution is similar to the one given in Example 4.4.

30. Solving for L = 61.5 gal and H = 38.5 gal.
31. After, 2 h, the concentration, c = 60/[(5)(2) + 1] = 5.45 g.
32. D = 930.2 kg/h, B = 1069.8 kg/h, w_B = 16.3 kg/h, w_{NB} = 90.7 kg/h.
33. Solution is done through MATLAB.
34. Solution is done through MATLAB.
35. Solution is done through Excel.
36. Composition of stream A is found by doing a T M B around column 1.
37. Solution is done through MATLAB.
38. Power required for pumping $= 0.0006 \times 133443.2/0.6$

$$= 133.4\,\text{W} = 133.4/736\,\text{HP} = 0.181\,\text{HP}.$$

40. A, is readily calculated = $Q/U\Delta T_{LM}$ = 540,000/[60 × 59.6] = 151 ft^2.
41. Q = 3.9 × 10^6 w/m^2.
42. Percentage loss of ethanol = 100 × (4.554 /430) = 1.06.
43. Composition of top product = $y_D = 69\%$ hexane.
44. D = 169.492 kmol/h, and the number of theoretical plates is estimated utilizing the McCabe–Thiele graphical construction method and found to be equal to 6.8. And the location of the feed plate is fifth plate (counting from the top).
45. The equilibrium data are plotted utilizing McCabe–Thiele method; the number of theoretical plates is found to be = 11 for the feed introduced at the eighth plate.
46. Q = $(T_s - T_a)/[R_1 + R_2]$ = (355 − 82.5)/1.8 = 151.4 Btu/h.
    Compare the Q lost for bare pipe to the one with insulation.
47. S = 4479.6 kg/h, that is, the steam required = 4479.6 kg/h.
48. Using MATLAB, F_1, F_2, and F_4 in mol/h are found to be x = 33,333.3333, 168,383.8384, and 3,333.3333, respectively.
49. The boiling point in the first effect will be = 206.5° and in the second effect = 181.5°.
50. Problem is attempted using MATLAB.

Appendix B: Mathematical Methods and Review Notes

B.1 USEFUL CONSTANTS

B.1.1 UNIVERSAL MATHEMATICAL CONSTANTS

A mathematical constant is a special number, usually a real number, which is *significantly interesting in some way* (Weisstein, 2011). Constants arise in many areas of mathematics, with constants such as e and π occurring in such diverse contexts as geometry, number theory, and calculus.

It is a diverse matter; what it means for a constant to arise *naturally* and what makes a constant *interesting* is ultimately a matter of taste, and some mathematical constants are more notable for historical reasons than for their intrinsic mathematical interest.

Zero and One (and i, and ...)	0 and 1 (and √(-1), and ...)
π, Archimedes' constant	3.141 592 653 589 793 238 462 643 •••
e, Euler number, **Napier's constant**	2.718 281 828 459 045 235 360 287 •••
γ, Euler-Mascheroni constant	0.577 215 664 901 532 860 606 512 •••
√2, Pythagora's constant	1.414 213 562 373 095 048 801 688 •••
Φ, Golden ratio	1.618 033 988 749 894 848 204 586 •••
φ, inverse golden ratio (often confused with Φ)	0.618 033 988 749 894 848 204 586 •••
δ_s, Silver ratio / mean	2.414 213 562 373 095 048 801 688 •••

Value	Description
3.14159265358979323846	Pi
2.71828182845904523 54	e
1.44269504088896340 74	log_2 e
0.43429448190325182765	log_10 e
0.69314718055994530942	log_e 2
2.30258509299404568402	log_e 10
1.57079632679489661923	pi/2
0.78539816339744830962	pi/4
0.31830988618379067154	1/pi

B.1.2 Values of Gas Constants (Jensen, 2003)

8.3144621(75) J/K · mole
0.08205746(14) L · atm/K · mole
1.9858775(34) cal/K · mole
8.3144621(75) L · kPa/K · mole
8.3144621(75) m^3 · Pa/K · mole
8.205746 × 10^{-5} m^3 · atm/K · mole
82.05746 cm^3 · atm/K · mole
62.36367(11) L · Torr/K · mole
6.132440(10) ft · lbf/K · g · mole
10.73159(2) ft^3 · psi/°R · lb · mole
0.7302413(12) ft^3 · atm/°R lb · mole
1.31443 ft^3 · atm/K · lb · mole
998.9701(17) ft^3 · mm Hg/K · lb · mole
1.986 Btu/lb · mole · °R

B.2 ROOT-FINDING METHODS

Root finding is the determination of solutions to single-variable equations or to systems of n equations in n unknowns (provided that such solutions exist). The basics of the method revolve around the determination of *roots*.

B.2.1 Introduction

Two approaches can be pursued to seek a solution: analytical and numerical.

An *analytical* solution to an equation or system is a solution that can be arrived at exactly using some mathematical tools. For example, consider the function $y = \ln(x)$, graphed in the following.

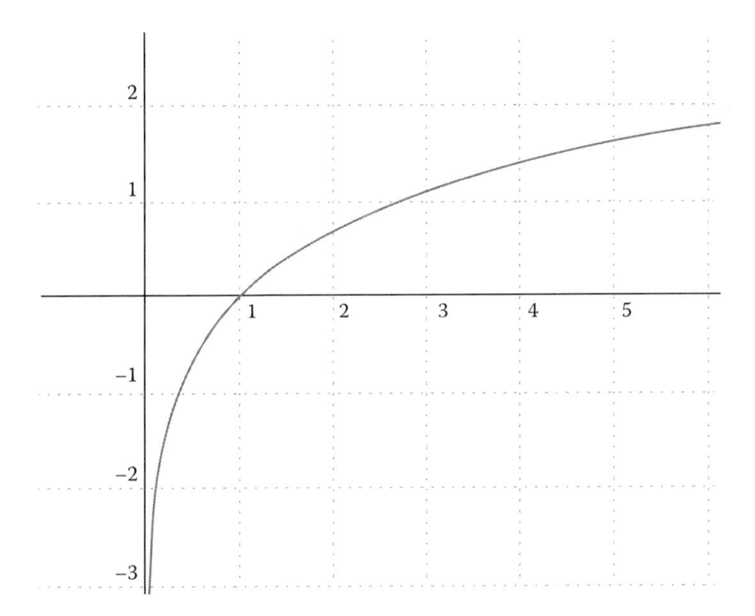

The root of this function is, by convention, when $y = 0$ or when this function crosses the x-axis. Hence, the root will occur when $\ln(x) = 0 \geq x = e^0 = 1$.

The answer $x = 1$ is an analytical solution because through the use of algebra, we were able to come up with an exact answer. On the other hand, attempting to solve other equations could be hard to achieve. In such cases, it is necessary to seek a *numerical* solution, in which *guesses* are made until the answer is *close enough*.

B.2.2 Numerical Root-Finding Methods–Algorithms

A root-finding algorithm is a numerical method, or algorithm, for finding a value x such that $f(x) = 0$, for a given function f. Such an x is called a root of the function.

Root finding methods are classified into:

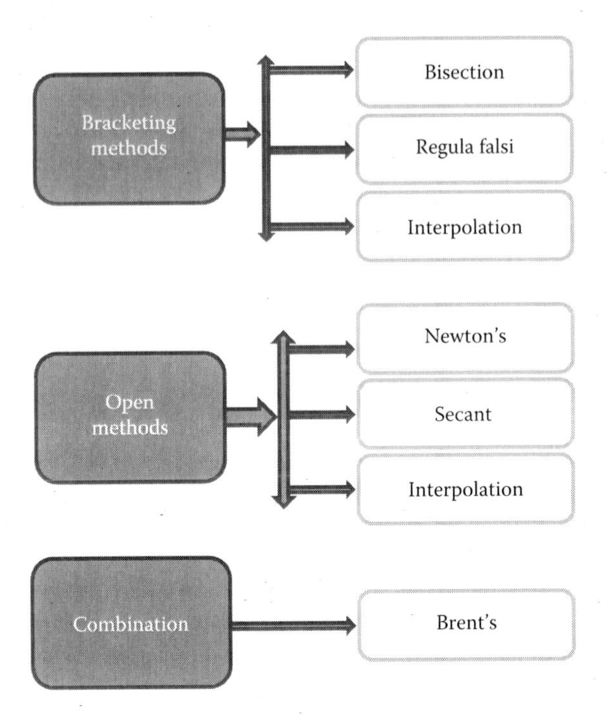

Three methods are described briefly:

1. Bisection method
2. Newton–Raphson method (Tangent method)
3. Secant method

Roots can be simple or multiple (p. 66)

x^* is a root of f having multiplicity q if

$$f(x) = (x - x^*)^q\, g(x) \text{ with } g(x^*) \neq 0$$

$$f(x^*) = f'(x^*) = \cdots = f^{(q-1)}(x^*) = 0 \text{ and } f^{(q)}(x^*) \neq 0$$

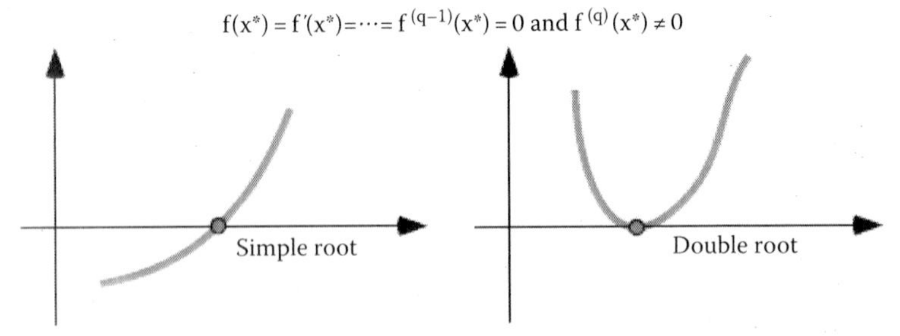

Use theory

All roots of $x - \cos x = 0$ lie in the interval $[-1,1]$

Proof:
$x = \cos x$
$\Rightarrow \quad |x| = |\cos x| \le 1$

Use graphics

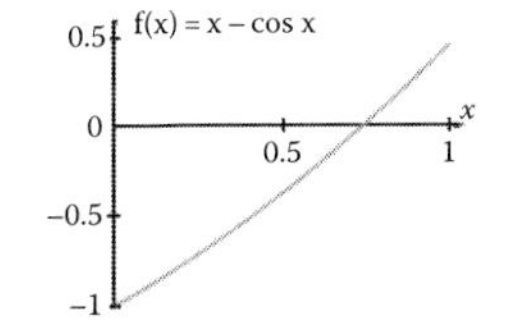

Three rules

1. Graph the function
2. Make a graph of the function
3. Make sure that you have made a graph of the function

Difficult cases:

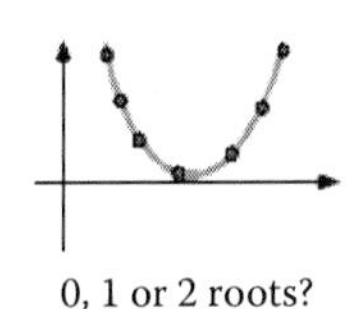

0, 1 or 2 roots?

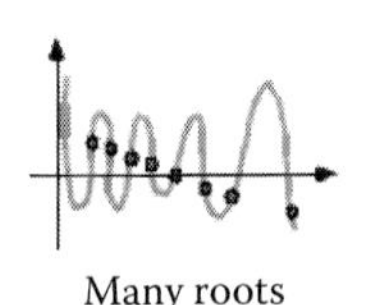

Many roots

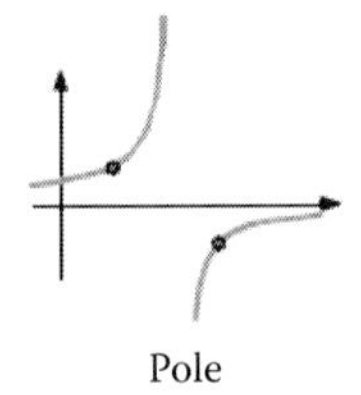

Pole

Bisection traps a root in a shrinking interval

Bracketing-interval theorem

If f is continuous on [a,b] and $f(a) \cdot f(b) < 0$ then f has at least one zero in (a,b).

Bisection method

Given a bracketing interval [a,b], compute $x = \frac{a+b}{2}$ and sign(f(x));

repeat using [a,x] or [x,b] as new bracketing interval.

```
function x=bisection(f,a,b,tol)
sfb=sign(f(b));
width=b-a;
disp('  a             b             sfx')
while width>tol
    width=width/2;
    x=a+width;
    sfx=sign(f(x));
    disp(sprintf('%0.8f   %0.8f   %2.0f',  [a b sfx]))
    if sfx==0, a=x; b=x; return
    elseif sfx == sfb, b=x;
    else, a = x; end
end
```

```
>> f=@(x) x-cos(x);
>>bisection(f,0.7,0.8,1e-3);

   a             b          sfx
0.70000000   0.80000000    1
0.70000000   0.75000000   -1
0.72500000   0.75000000   -1
0.73750000   0.75000000    1
0.73750000   0.74375000    1
0.73750000   0.74062500   -1
0.73906250   0.74062500    1
```

Bisection is slow but dependable.

Advantages

- Guaranteed convergence
- Predictable convergence rate
- Rigorous error bound

Disadvantages

- May converge to a pole
- Needs bracketing interval
- Slow

Newton – Raphson method uses the tangent

Iteration formula

$$x_{k+1} = x_k - \frac{f(x_k)}{f'(x_k)}$$

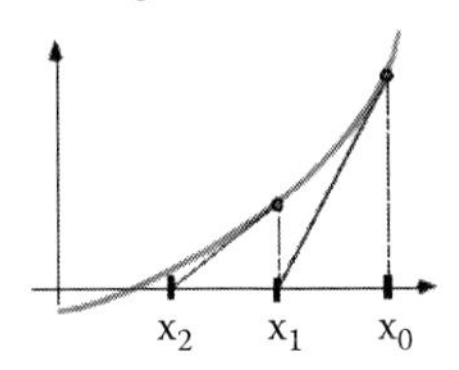


```
function x=newtonraphson(f,df,x,nk)
disp('k       x_k               f(x_k)             f''(x_k)          dx')
for k=0:nk
    dx=df(x)\f(x);
    disp(sprintf('%d      %0.12f        %9.2e   %1.5f    %15.12f',[k,x,f(x),df(x),dx]))
    x = x - dx;
end
```

```
>> f=@(x) x-cos(x); df=@(x) 1+sin(x);
>>newtonraphson(f,df,0.7,3);
k     x_k              f(x_k)      f'(x_k)     dx
0   0.700000000000   -6.48e-02   1.64422  -0.039436497848
1   0.739436497848    5.88e-04   1.67387   0.000351337383
2   0.739085160465    4.56e-08   1.67361   0.000000027250
3   0.739085133215    2.22e-16   1.67361   0.000000000000
```

Secant method is derivative-free

Iteration formula

$$x_{k+1} = x_k - \frac{f(x_k)}{\left(\dfrac{f(x_k) - f(x_{k-1})}{x_k - x_{k-1}}\right)}$$

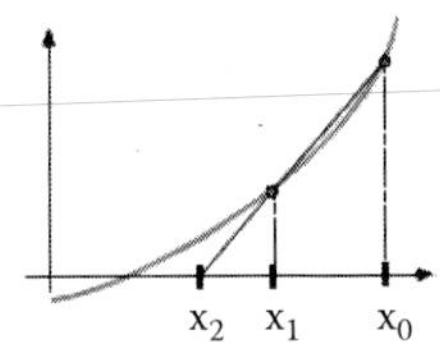


```
function xx = secant(f,xx,nk)
disp('k x_k             f(x_k)')
ff = [f(xx(1)), f(xx(2))];
h = 10*sqrt(eps);
for k = 0:nk
  disp(sprintf('%d %17.14f %14.5e',...
       [k,xx(1),ff(1)]))
  if abs(diff(xx))>h
    df = diff(ff)/diff(xx);
  else
    df = (f(xx(2)+h)-ff(2))/h;
  end
  xx = [xx(2),  xx(2)-ff(2)/df];  % update xx
  ff = [ff(2), f(xx(2))];         % update ff
end
```

```
>> f = @(x)  x-cos(x);
>> secant(f, [0.7 0.8], 6);

k    x_k                  f(x_k)
0  0.70000000000000  -6.48422e-02
1  0.80000000000000   1.03293e-01
2  0.73856544025090  -8.69665e-04
3  0.73907836214467  -1.13321e-05
4  0.73908513399236   1.30073e-09
5  0.73908513321516  -1.77636e-15
6  0.73908513321516   0.00000e+00
```

B.3 TIPS ON SOLVING PROBLEMS

B.3.1 Mathematical Modeling

The basic problems that we face in sciences and engineering, as far as mathematical modeling is concerned, fall into three main categories:

1. *Equilibrium problems*: This type of problems is recognized as steady state, where a solution does not change with time.
2. *Eigenvalue problems*: This type of problems is recognized as extensions of equilibrium problems in which critical values of certain parameters are to be determined in addition to the corresponding steady-state configuration.
3. *Transient, time-varying, or propagation problems*: This type of problems is concerned with predicting the subsequent behavior of a system from the knowledge of the initial stage.

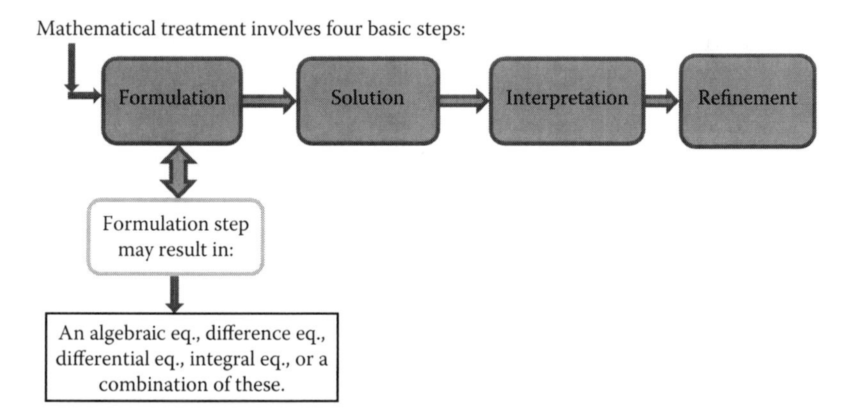

B.3.2 The Role of Synthesis and Analysis

In handling problems in the chemical process industry, two stages are usually involved: synthesis and analysis. Synthesis is the development and the creation of a new idea, as demonstrated in process design in Chapter 9. Analysis, on the other hand, confirms or rejects new ideas based on a feasibility study by subjecting them to careful scrutiny to see if they obey or conflict the laws of physics and chemistry.

As explained earlier, the analysis process comprises the formulation process and the development of the various equations, and their subsequent solutions for the unknown variables.

REFERENCE

Jensen, W.B. The universal gas constant R, *Journal of Chemical Education* 80(7): 731, July 2003.

Weisstein, E.W., *Constant*, Wolfram MathWorld. http://mathworld.wolfram.com/Constant.html. Retrieved April 13, 2011.

Appendix C: Conversion Tables

Common Engineering Design Conversion Factors

Given	Equivalent Value	Unit of Measure
Length [L]		
Foot (ft)	0.304800	Meter (m)
Inch (in.)	25.4000	Millimeter (mm)
Mile (mi)	1.609344	Kilometer (km)
Area $[L]^2$		
ft^2	0.092903	m^2
$in.^2$	645.16	mm^2
$in.^2$	6.45160	cm^2
Volume $[L]^3$ and capacity		
$in.^3$	16.3871	cm^3
ft^3	0.028317	m^3
ft^3	7.4805	Gallon
ft^3	28.3168	Liter (L)
Gallon	3.785412	Liter
Energy, work, or heat $[M] [L]^2 [t]^{-2}$		
Btu	1.05435	kJ
Btu	0.251996	kcal
Calories (cal)	4.184*	Joules (J)
ft · lbf	1.355818	J
ft · lbf	0.138255	kgf · m
hp · h	2.6845	MJ
kWh	3.600	MJ
m · kgf	9.80665*	J
N · m	1	J
Flow rate $[L]^3 [t]^{-1}$		
ft^3/min	7.4805	gal/min
ft^3/min	0.471934	l/s
gal/min	0.063090	l/s
Force or weight $[M] [L] [t]^{-2}$		
kgf	9.80665*	Newton (N)
lbf	4.44822	N
lbf	0.453592	kgf

(*Continued*)

Common Engineering Design Conversion Factors

Given	Equivalent Value	Unit of Measure
Fracture toughness		
ksi · sqrt(in.)	1.098800	MPa sqr(m)
Heat content		
Btu/lbm	0.555556	cal/g
Btu/lbm	2.324444	J/g
Btu/ft^3	0.037234	MJ/m^3
Heat flux		
Btu/h · ft^2	7.5346E−5	cal/s · cm^2
Btu/h · ft^2	3.1525	W/m^2
cal/s · cm^2	4.184*	W/cm^2
Mass density [M] [L]$^{-3}$		
lbm/in.^{3}	27.68	g/cm^3
lbm/ft^3	16.0184	kg/m^3
Power [M] [L]^2 [t]$^{-3}$		
Btu/h	0.292875	Watt (W)
ft · lbf/s	1.355818	W
Horsepower (hp)	745.6999	W
Horsepower	550*	ft · lbf/s
Pressure (fluid) [M] [L]$^{-1}$ [t]$^{-2}$		
Atmosphere (atm)	14.696	lbf/in.^{2}
atm	1.01325E5*	Pascal (Pa)
lbf/ft^2	47.88026	Pa
lbf/in.^{2}	27.6807	in. H_2O at 39.2°F
Stress [M] [L]$^{-1}$ [t]$^{-2}$		
kgf/cm^2	9.80665E−2*	MPa
ksi	6.89476	MPa
N/mm^2	1	MPa
kgf/mm^2	1.42231	ksi
Specific heat		
Btu/lbm · °F	1	cal/g · °C
Temperature		
Fahrenheit	(°F − 32)/1.8	Celsius
Fahrenheit	°F + 459.67	Rankine
Celsius	°C + 273.16	Kelvin
Rankine	R/1.8	Kelvin
Thermal conductivity		
Btu · ft/h · ft^2 · °F	14.8816	cal · cm/h · cm^2 · °C

C.1 ENERGY UNITS AND CONVERSIONS

1 Joule (J) is the MKS unit of energy, equal to the force of 1 N acting through 1 m.
1 Watt is the power of a Joule of energy per second.

$$\text{Power} = \text{current} \times \text{voltage } (P = IV).$$

1 Watt is the power from a current of 1 A flowing through 1 V.
1 kilowatt is a 1000 W.
1 kilowatt-hour (kWh) is the energy of 1 kW power flowing for 1 h ($E = Pt$).

$$1 \text{ kWh} = 3.6 \times 10^6 \quad \text{or} \quad J = 3.6 \text{ million Joules.}$$

1 cal of heat is the amount needed to raise 1 g of water 1°C.

$$1 \text{ calorie (cal)} = 4.184 \text{ J.}$$

(The calories in food ratings are actually kilocalories.)
A British thermal unit (Btu) is the amount of heat necessary to raise 1 lb of water by 1 degree Fahrenheit (°F).

1 Btu = 1055 J (the mechanical equivalent of heat relation).

1 Btu = 252 cal = 1.055 kJ.

1 Quad = 10^{15} Btu (world energy usage is about 300 Quads/year, and United States is about 100 Quads/year in 1996).
1 therm = 100,000 Btu.
1000 kWh = 3.41 million Btu.

C.2 POWER CONVERSION

1 horsepower (hp) = 745.7 Watts

C.3 GAS VOLUME TO ENERGY CONVERSION

One thousand cubic feet of gas (Mcf) → 1.027 million Btu = 1.083 billion J = 301 kWh

1 therm = 100,000 Btu = 105.5 million J = 29.3 kWh
1 Mcf → 10.27 therms

C.4 ENERGY CONTENT OF FUELS

Coal 25 million Btu/ton
Crude oil 5.6 million Btu/barrel
Oil 5.78 million Btu/barrel = 1700 kWh/barrel
Gasoline 5.6 million Btu/barrel (a barrel is 42 gallons) = 1.33 therms/gallon
Natural gas liquids 4.2 million Btu/barrel
Natural gas 1030 Btu/ft^3
Wood 20 million Btu/cord

C.5 CO_2 POLLUTION OF FOSSIL FUELS

Pounds of CO_2 per billion BTU of energy:
Coal 208,000 lb
Oil 164,000 lb
Natural gas 117,000 lb

Ratios of CO_2 pollution:
Oil/natural gas = 1.40
Coal/natural gas = 1.78

Pounds of CO_2 per 1000 kWh at 100% efficiency:
Coal 709 lb
Oil 559 lb
Natural gas 399 lb

Bibliography

Abdel-Aal, H.K., Aggour, M.A., and Fahim, M.A., *2-Petroleum and Gas Field Processing*, 2nd edn., Chemical Industries, London, U.K.: CRC Press/Taylor & Francis Group, 2016.

Abdel-Aal, H.K. and Alsahlawi, M.A. (Eds.), *Petroleum Economics and Engineering*, 3rd edn., London, U.K.: CRC Press/Taylor & Francis Group, 2014.

Abdel-Aal, H.K., Ibrahim, A.A., Shalabi, M.A., and Al-Harbi, D.K., Dual-purpose chemical desalination process, *Desalination* 113: 19–25, 1997.

Allchin, F.R., India: The ancient home of distillation?, *Man* 14(1): 55–63, 1979.

Association for the Advancement of Modeling and Simulation Techniques in Enterprises (AMSE).

Avallone, E.A. and Baumeister, T. (Eds.), *Marks' Standard Handbook for Mechanical Engineers*, 11th edn., New York: McGraw-Hill, Inc., 1997.

Baasal, W.D., *Preliminary Chemical Engineering Plant Design Hardcover*, 2nd edn., November 30, 1989.

Bird, R.B., Stewart, W.E., and Lightfoot, E.N., *Transport Phenomena*, revised 2nd edn., New York: John Wiley & Sons, 2007.

Brief history of the SI. BIPM. Retrieved April 21, 2009. The International Bureau of Weights and Measures. /wiki/French_language" French: *Bureau international des poids et mesures "BIPM."*

Cavalcade Publishing, The Cavalcade o' chemistry, 2001, https://chemfiesta.org/. Retrieved on June 10, 2012.

Chemical Engineering Learning Resource, mmsubbu, 1953, http://www.msubbu.in, Accessed on September 12, 2014.

Cutlip, M.B., *Problem Solving in Chemical Engineering with Numerical Methods*, Upper Saddle River, NJ: Prentice Hall, 1999.

Cutlip, M.B., et al., A collection of 10 numerical problems in chemical engineering, ftp://folklore.org.il/shacham/publ_papers/CAEE_6_169_98.pdf.

Deen, W.M., *Analysis of Transport Phenomena*, Oxford, U.K.: Oxford University Press, 1998.

DePriester, C.L., *Chemical Engineering Progress on Symposium Series* 7(49): 1–43, 1953.

Dimian, A.C., Bildea, C.S., and Kis, A.A., *Integrated Design and Simulation of Chemical Processes*.

Energy Institute, Resources, https:// www.schoolscience.co.uk, retrieved on April 20, 2014.

Felder, R.M. and Rousseau, R.W., *Elementary Principles of Chemical Processes*, 3rd edn., New York: John Wiley & Sons, Inc., 2005.

Finlayson, B.A., *Introduction to Chemical Engineering Computing*, 2nd edn. (update edition), John Wiley & Sons, 2014.

Forbes, R.J., *A Short History of the Art of Distillation: From the Beginnings Up to the Death of Cellier Blumenthal*. Leiden, the Netherlands: Brill, 1970, pp. 57, 89. Retrieved on June 29, 2010.

Gani, R., *Computer Aided Chemical Engineering*, Lyngby, Denmark: CAPEC, Department of Chemical and Biochemical Engineering, Technical University of Denmark.

Geankoplis, C.J., *Transport Processes and Separation Process Principles*, 4th edn., Engelwood Cliffs, NJ: Prentice Hall, 2003.

Griskey, R., *Transport Phenomena and Unit Operations—Combined Approach*, New York: John Wiley, 2002.

Griskey, R.G., *Transport Phenomena and Unit Operations: A Combined Approach* (paperback), New York: John Wiley, 2006.

Harj, B., *Cambridge Reactor Design Limited*, Cambridge, U.K.: Cambridge University.

Kunesh, J.G., Lahm, L., and Yanagi, T., Commercial scale experiments that provide insight on packed tower distributors, *Industrial and Engineering Chemistry Research* 26(9): 1845, 1987.

Long, C. and Sayma, N., *Heat Transfer: Exercises*, 1st edn., Brighton, UK: Ventus Publishing ApS, 2010. http://www.leka.lt/sites/default/files/dokumentai/heat-transfer.pdf. Retrieved June 12, 2013.

Mellon, C., MPEC problem formulations in chemical engineering applications, http://numero.cheme.cmu.edu/uploads/rp.pdf, Chemical Engineering Department. Retrieved on June 16, 2007.

Metcalf & Eddy, *Wastewater Engineering: Treatment and Reuse*, 4th edn., New York: McGraw-Hill, 2003, p. 1449.

Misterguch, The six types of chemical reactions, 2012, http://misterguch.brinkster.net/6typesofchemicalrxn.html.

Moore, F. and Rukovena, F., Random packing, vapor and liquid distribution: Liquid and gas distribution in commercial packed towers, *Chemical Plants and Processing* 11–15, August 1987.

Needham, J., *Science and Civilisation in China*, Cambridge, U.K.: Cambridge University Press, 1980.

Nonweiler, T.R., *Computational Mathematics: An Introduction to Numerical Approximation*, New York: John Wiley & Sons, 1986.

Othmer, D.F., Distillation—Some steps in its development, in W.F. Furter (ed.), *A Century of Chemical Engineering*; Separation of liquid–liquid mixtures (solutions), IUPAC, 1982.

Palm, W.J., *Introduction to MATLAB (R) for Engineers*, 3rd edn., New York: McGraw-Hill, 2011.

Perry, R.H. and Green, D.W., *Perry's Chemical Engineers' Handbook*, 6th edn., New York: McGraw-Hill, 1984.

Perry, R.H. and Green, D.W. (Eds.), *Perry's Chemical Engineers' Handbook*, 7th edn., New York: McGraw-Hill, 1997.

Plawsky, J.L., *Transport Phenomena Fundamentals*, New York: CRC Press, April 2001, pp. 1–3.

Press, W.H. et al., *Numerical Recipes in C: The Art of Scientific Computing*, 3rd edn., Cambridge, U.K.: Cambridge University Press, 2007.

Rice, R.G. and Do, D.D., *Applied Mathematics and Modeling for Chemical Engineers*, New York: John Wiley & Sons, Inc., 1995.

Roy, G.K., *Solved Examples in Chemical Engineering*, Delhi, India: Khanna Publishers, 2010.

Schaschke, C.J., *Solved Practical Problems in Fluid Mechanics*, Dundee, Scotland: Abertay University.

Seader, J.D. and Henley, E.J., *Separation Process Principles*, New York: Wiley, 1998.

Shalabi, M.A., Harji, B.H., and Kenney, C.N., *Journal: Cheminform* 27(28), 2010.

Spiegel, L., A new method to assess liquid distributor quality, *Chemical Engineering and Processing* 45(11): 1011, 2006.

Study Guides and Strategies, *Welcome to Problem Solving with Algorithms and Data Structures*, n.d., http://www.studygs.net/scimethod.htm, retrieved on December 4, 2014.

Subramanian, M., *Process Calculations—Solved Problems in Chemical Engineering*, n.d., http://www.msubbu.in/sp/pc/, retrieved on September 13, 2014.

Tampere University of Technology, tut.fi/~piche/numa/lecture0506.pdf.

Theodore, L. and Ricci, F., *Mass Transfer Operations for the Practicing Engineer*, Hoboken, NJ: John Wiley & Sons, 2010.

Tyler, G., Hicks, P.E., and Chopey, N.P., Heat transfer, in *Handbook of Chemical Engineering Calculations*, 4th edn., New York: McGraw-Hill Professional, 2012.

Wikibooks, *Introduction to Chemical Engineering.*

Yang, X.S., *Introduction to Computational Mathematics*, Singapore: World Scientific Publishing, 2008.

Zondervan, E., *A Numerical Primer for the Chemical Engineer*, Boca Raton, FL: CRC Press, 2015.

Index

A

Absolute pressure
  example, 18
  gauge pressure and, 74–75
Absorption
  analytical methods, 154–156
  concepts, 148–149
  defined, 119, 147
  design approach, 149
  example, 152–154
  graphical methods, 150
  material balance calculations, 151
  role in natural gas field processing, 156–157
  transfer unit concept for, 156
Absorption factor method, 155
Acetonitrile, Pxy diagram of, 64–65
Acid–base reaction, 168
Acid concentration, finding, 41–42
Algorithms, 36
  solved examples, 38
Alternatives to storage of ammonia (case study), 202–204
Amagat's law, 13
American engineering system, 5
American Petroleum Institute (API) process pumps, 85
Analysis process, 241
Analytical methods, 154–156
Annual rate of return, 195
Annuity, 194–195
Antoine coefficients, 51
Antoine equation, 51
Applications of differential equations (case study), 216–218
Arrhenius equation, 169
Association for the Advancement of Modeling and Simulation Techniques in Enterprises (AMSE), 48
Avogadro's hypothesis, 14
Axial flow pumps, 85
A–Z chart for plant design, 190–191

B

Bernoulli's equation, 77, 87
  flow direction, determination, 90
  head form of, 78
Binary distillation, 134–135
Bisection method, 239–240
Blackbody radiation, 107
Black box models, *see* Empirical models
Boiler feed pumps, 85
Bottom product, 124
Boyle's law, 9
  example, 15, 28
Bubble-point (BP) temperature
  defined, 24, 50
  determination, 51–52
  Excel, calculating by, 53
  Newton's method, calculating by, 52–53
  Raoult's law, 55–56

C

Cash recovery period, *see* Payout period
Catalytic reactor, 174–175
Celsius temperature scale, 6
Centigrade temperature scale, *see* Celsius temperature scale
Centimeter–gram–second (cgs) system, 3, 5
Centrifugal force, 83
Charles' law, 9
  example, 15, 28
Chemical desalting of brines (case study)
  consumption–production analysis, 212–214
  discussion and problem formulation, 215–216
  methodology, 210–211
  primary reaction, 211
  product formation reaction, 212
  secondary reaction, 212
  Solvay process and proposed process, 213
  synthesis process with modifications, 212
Chemical engineering
  defined, xxvii
  plant design, 188–189
  problems, solution, 45
  unit operation, 19–20
  unit process, 20
Chemical industry, 188
Chemical kinetics, *see* Reaction kinetics
Chemical plant, 21–22
  engineering aspect of, 48
Chemical plant, design of
  A–Z chart for, 190–191
  chemical engineering, 188–189
  economics role (*see* Economics in plant design)
  flow sheeting and flow diagram, 191
  other aspects, 191–192
  proposed procedure, 189–190

safety factors, 191, 192
scale-up practice, 191
Chemical process engineers, 21
Chemical process industry (CPI), 20–21
Chemical product engineers, 21
Chemical pumps, 85
Chemical reactions
categories of, 27, 167–168
equilibrium constant for, 24
free energy of formation in, 180–181
Chemical reactors
catalytic reactor, 174–175
chemical equilibrium, using MATLAB®, 184
design, 175–178
laboratory and industrial catalytic reactors, 173
models, classified, 172
Circulator pumps, 85
Combustion of heptane, 30
Combustion reaction, 167
Common engineering design conversion factors, 243–244
Compressible fluid, 70
Computational science, components, xxii
Computers, problem solving using, 46–47
Conceptual model, 49
Conduction, heat transfer by
Fourier's law, 97–98
solved examples, 99–101
thermal conductivity, 96, 98
through hollow cylinder, 99
through solids in series, 99
Consistency, 71
Consumption–production analysis, 212–214
Continuous stirred-tank reactor model (CSTR), 174
Convective heat transfer
convective heat transfer coefficient, 102
example, 103
fluid motion, 96
methodology of calculation, 102
natural convection *vs.* forced convection, 101
Conversions
energy units and, 245
gas volume to energy, 245
physical events, 24–25
power, 245
Countercurrent flow, 107, 108
CPI, *see* Chemical process industry
Cross flow, 107
Crude oil distillation, flow diagram for, xxiv

D

Dalton's law, 9–10, 50
Decomposition reaction, 167
Derived units, 3
Design problem in distillation, 130
Design variables for sizing piping, 23–24
Desorption, 119
Dewatering process, 85
Dew-point (DP) temperature
defined, 52
examples, 54, 56–57
Differential equations, 216–218
Differential rate law, 165, 168–170
Differential-stage contactors, 123
Differentiation, example, 39
Diffusional mass transfer, 20, 119
Diffusional unit operations, 21
Discounted cash-flow rate of return (DCFR), 196
solved examples, 199–200
Distillation
classification methods, 122
column diameter, 144
concepts and principles, 120–123
defined, 119
design problem, 130
flash distillation, 144–147
McCabe–Thiele method (*see* McCabe–Thiele method)
overhead condensers, 125–126
physical models for, 131
pillars for, 123–124
rating (operational) problem, 130–131
reactive, 147
reboilers, 127
reflux and reflux ratio, 126–127
shortcut methods (*see* Shortcut calculation methods)
units, features of, 124
Distillation columns
design of, 129–131
example, 58
number of trays, calculation of, 131
rectification and stripping sections in, 127, 130
types, 125, 126, 128
Distribution constant, 24, 147
Double displacement reaction, 168
DP temperature, *see* Dew-point temperature

E

Economic balance in piping, 80–81
Economics in plant design
annuity, 194–195
FCI and total capital investment, 193–194
profitability analysis, 195–196
Effective equilibrium curve (EEC), 139–140
Eigenvalue problems, 47, 241
Electromagnetic radiation, 106
Elementary numerical methods, applications using, 37–38

Empirical models, xxii
Energy balance equation
  Bernoulli's equation, 77–78
  general, 75–76
  mechanical, 76
  pressure drop and friction losses, 78–79
Energy balance, role of, 22
Energy content of fuels, 246
Energy conversion, gas volume to, 245
Energy units and conversions, 245
Engineering aspect of chemical process, 48
Engineering science, xxv
Entropy, 179–180
  solved examples, 181–182
Equations of state, 178–179
Equilibrium constant, for chemical reaction, 24
Equilibrium problems, 47, 241
Equilibrium stage, 122, 140
Equilibrium state, 178
Ethanol–water system, Txy diagram for, 59
Excel
  BP temperature calculation, 53
  flow direction, determination, 90–91
  McCabe–Thiele method, problem solving by, 134–135
  plate-to-plate calculations by, 137
  problem solving using, 37, 46
  shell and tube heat exchanger, calculation, 112–113
  solution of linear equation system, 40–41
  Txy diagram for benzene–toluene mixture using, 60–62, 64

F

Fahrenheit temperature scale, 6
False position (FP) method, 129
FCI, *see* Fixed capital investment
Fenske equation, 142
Fenske–Underwood–Gilliland, *see* Shortcut calculation methods
Fertile water, 210
Finite-stage contactors, 123
Fire pumps, 85
First law of thermodynamics, 26, 179
First-order reaction, 169, 170
Fixed-bed reactor, 176
Fixed capital investment (FCI), 193–194
Flash dewatering of raw sewage effluents (case study)
  conclusions, 207
  discussions, 208
  process description, 206–207
  proposed scheme, 204, 206
Flash distillation, 50, 144–145
  example, 58–59
  problem, 24, 25
Flash equation, 146–147
Flashing, 119
Flow arrangement, 107
Flow diagram, types, 191
Flow sheeting, 191
Flow work, 76
Fluid dynamics, xxv, 70
Fluid flow, 20
  classification of, 71–72
  in pipes, 83
Fluid mechanics, definitions and terminology in, 70–71
Fluids
  defined, 69
  nature of, 71
Fluid statics, 70
  gauge pressure and absolute pressure, 74–75
Forced convection, natural convection *vs.*, 101
Forces, types of, 70
Formulating mathematical modeling, 47–49
Fossil fuels, CO_2 pollution of, 246
Fourier's law, 96–98
Fractional distillation, *see* Distillation
Free energy of formation of reaction, 180–181
Free stream velocity, 74
Friction losses, pressure drop and, 78–79
Fuels, energy content of, 246

G

Gas constant, 10
  values of, 236
Gas laws
  Amagat's law, 13
  Avogadro's hypothesis, 14
  Boyle's law, 9
  Charles' law, 9
  Dalton's law, 9–10
  ideal gas law, 10–12
  Raoult's law, 12–13
Gas volume to energy conversion, 245
Gauge pressure, and absolute pressure, 74–75
General energy balance equation, 75–76
Gilliland method, 142, 143
Gravity force, 70
Gray box models, *see* Semiempirical models

H

Heat balance (HB), 25, 188
Heat exchangers
  design, 109–110
  example, 110
  LMTD, 108–109
  shell and tube, 109
  types, 107–108

Heat flow, 95
Heat transfer
  coefficient, calculating, 14
  by conduction (*see* Conduction, heat transfer by)
  by convection, 101–103
  modes of, 96–97
  by radiation, 106–107
  temperature and heat flow, 95
Henry's law, 13
Heterogeneous catalysis reactor types, 175
High-pressure pumps, 85
Hollow cylinder, conduction through, 99
Homemade soap, 216
Horsepower (HP) for pump
  calculation, 86
  solved examples, 89

I

Ideal fluid flow, 72
Ideal gas constant, *see* Gas constant
Ideal gas law, 10–12
Industrial catalytic reactors, 173
Industrial production of CO_2, 29
Industrial pumps, 85
Integrated rate law, 165, 170–172
Integration, example, 38–39
Internal-flow convection correlations, 116–117
International System of Units, 3–5
Interphase operations, 119, 120
Intraphase separation, 121
Irrotational flow, *see* Ideal fluid flow

K

Kelvin temperature scale, 6
Kettle-type reboiler, 127
Kinematic viscosity, 71
Kinetic energy
  of flowing water, 15
  of missile, 30
Kremser equations, 155

L

Laboratory catalytic reactors, types, 173
Laminar flow
  friction factor, 79
  parameters, 73–74
Linear equation system
  Excel, solution by, 40–41
  MATLAB®, solution using, 54–55
  numerical solution of, 36
Linear programming problem, 37
Liquid–gas mass transfer operations, 119
Liquid product, 127
Lithium hydroxide, 31
Log mean temperature difference (LMTD), 108–109
  calculating, 110

M

Magnesium chloride, 210
Magnesium extraction from seawater (case study), 208–210
Magnesium metal, production of, 210
Marine pumps, 85
Mass and energy conservation, 25
Mass transfer, xxv, 96, 122, 138
Material balance (MB)
  calculations, 25, 151, 188
  in process design, 190
  role, 22
Mathematical constant, 235–236
Mathematical formula, principles to, 24–25
Mathematical modeling, 241
  formulating, 47–49
MATLAB®
  chemical equilibrium inside chemical reactor, 184
  derivative function, calculating, 39
  distillation column, 158
  heat transfer rate, calculations by, 100
  linear equation system using, 54–55, 111
  problem solving, 37, 46
Matrix, solution using, 39, 42–43
McCabe–Thiele method
  assumptions, 132
  plate-to-plate calculations, 135–137
  problem solving by using Excel, 134–135
  procedure, 132–134
  stage design and efficiency, 140–141
  total and minimum reflux, 138
  tray efficiency, 138–140
Mean residence time, 176–177
Mechanical energy balance, 76
Methanol in reactor, per pass conversion of, 182–183
Methyl acetate, production of, 218, 219
Minimum number of stages, 129
Minimum reflux (MR), 129, 138
Mixed flow pumps, 85
Modern chemical engineering, xxiii
Molar constant, *see* Gas constant
Molar mass, 11
Momentum, heat, and mass (MHM), xxiv
Moody chart, 79
MSFE, *see* Multistage flash evaporation
Mud pumps, 85
Multistage flash evaporation (MSFE), 204, 206, 208
Murphree efficiency, 139

N

Natural convection *vs.* forced convection, 101
Natural gas field processing, absorption role in, 156–157
Nature of fluids, 71
Nernst equation, 10
Net present value (NPV), 196
  example, 200
Newtonian fluid, 71
Newton–Raphson method, 240
  example, 38
Newton's law of viscosity, xxvi–xxvii
Newton's method, BP temperature calculation by, 52–53
Newton's second law, 70–71
Noncompressible fluid, 70
Nonlinear equations, numerical solution of systems of, 36–37
Nonnewtonian fluid, 71
Nonuniform flow, 72
Normal stress, 70
Number of trays, calculation of, 131
Numerical methods
  applications using elementary, 37–38
  areas and domain of, 36–37
  development, 35
  formulation of mathematical model through, 49
Numerical root-finding methods, 237–240

O

One-dimensional steady-state heat conduction, 97, 98
Operating pressure
  effect of, 127–128
  solved example, 128–129
Optimum economic pipe diameter, 80–81
Ordinary differential equations, 216
Overall efficiency, 139
Overall heat transfer coefficient, 103–105
  example, 105–106
Overhead condensers, 125–126
Overhead product, 124, 127

P

Parallel flow, 107, 108
Partial condenser, 125–126
Partial reboilers, 127
Parts-per notation, 26
Pascal, 7
Payback time, *see* Payout period
Payout period, 196
Petrochemical pumps, 85
Petroleum fractionation process, xxiii–xxv
Physical changes, *see* Physical separation
Physical events, conversion of, 24–25
Physical models for distillation, 131
Physical operations, xxiii
Physical separation, 20
  process, 30
Physicochemical changes, 21, 188
Pinch point, 138
Pipe diameter and pressure drops, 82–83
Pipeline capacity, 83
Piping
  economic balance in, 80–81
  pipe diameter and pressure drops, 82–83
  sizing, 23–24, 80
  wall thickness, procedure to calculate, 81–82
Plant design, *see* Chemical plant, design of
Plate-to-plate calculations
  rectification column, 135–137
  stripping column, 137
Pneumatic pumps, 85
Polonium, radioactive decay of, 42
Potential flow, *see* Ideal fluid flow
Power conversion, 245
Practical component, xxii
Prandtl numbers, 116
Preferential salt separation (PSS), 209
Present value index (PVI), 196
Pressure, 70
Pressure drop
  and friction losses, 78–79
  pipe diameter and, 82–83
Pressure force, 70
Pressure pumps, 85
Pressure scale, 7, 9
Problem solving using computers, 46–47
Process design, 22
Process equipment, 23–24
Process pumps, 85
Process variables
  pressure scale, 7, 9
  temperature scales, 6–8
Product formation reaction, 212
Profitability analysis using economic indicators
  annual rate of return, 195
  DCFR and PVI, 196
  NPV, 196
  payout period, 196

Programming languages, 36
Propagation problems, 47, 241
Pumps
classification and types of, 84
fluid flow, 83
horsepower, calculation of, 86
powerful, 80
selection, 85–86
Pxy diagram of acetonitrile, 64–65

Q

Quantitative process, 48

R

Radiation, heat transfer by, 97, 106–107
Radioactive decay of polonium, 42
Raoult's law, 12–13, 50, 55, 63
Rate constant, 169
Rate law, forms of
differential rate law, 165, 168–170
integrated rate law, 170–172
Rate of shear stress, 73–74
Rating (operational) problem, in distillation, 130–131
Reaction kinetics, 166
chemical reactions, categories of, 167–168
reaction rates (*see* Reaction rates)
role, 27
Reaction rates
differential rate law, 165, 168–170
integrated rate law, 170–172
Reactive distillation, 147
case study, 218–219
Real fluid, 71
flow, 72
Reboilers, 127
Rectification, 119
in distillation columns, 127, 130
Reflux, 126–127
Reflux ratio, 126–127, 129, 139
Reforming unit, xxiv
Relative volatility, 121
Reynolds (Re) number, 79, 116, 117
Root-finding methods, 236
analytical solution, 237
numerical, 237–240
Rule-of-thumb economic velocity, 83

S

Safety factors in plant design, 191, 192
Saponification process, 216, 217
Scientific point of view, 48
Scientific problems, solving, xxii
Secant method, 240
Second law of thermodynamics, 26, 179
Second-order reaction, 169, 170
Semiempirical models, xxii
Separation factor (SF), 122
Sewage treatment process, 205
Shear force, 70
Shear stress, 70, 74
Shell and tube heat exchanger, 109
solved example, 112–113
Shortcut calculation methods
column diameter, 144
Fenske equation, 142
Gilliland method, 142, 143
Underwood equation, 142–143
Single displacement reaction, 167
Single-stage contact unit, 140
Sinking-fund factor (SFF), 194
Slurry pumps, 85
Soap, 216
Sodium chloride, 216, 217
Sodium sulfite, 27–28
Solar hydrogen, to desert development using ammonia, 202
Solar pumps, 86
Solids in series, conduction through, 99
Solvay process, 210
proposed process and, 213
Solved examples
absolute pressure, 18
absorption, 152–154
acetonitrile, Pxy diagram of, 64–65
acetylene and hydrogen, yield of, 184–186
acid concentration, finding, 41–42
algorithm, 38
annual depreciation costs, 198–199
bed temperature, estimation, 113–116
benzene, 31–32
benzene–toluene mixture, Txy diagram for, 60–62, 64
Bernoulli's equation, 87, 88
BP temperature calculation, 52–53, 55–56
chemical equilibrium inside chemical reactors, using MATLAB®, 184
combustion of heptane, 30
composition of streams, calculation, 159–160
conduction, heat transfer by, 99–101
DCFR, 199–200
dew-pressure calculation, 54, 56–57
diameter column, 163–164
differentiation, 39
distillation column, 58, 157
entropy, calculations, 181–182
ethanol–water system, Txy diagram for, 59

flash distillation (separation), 58–59
flow direction, determination, 89–91
flow rates for product streams, 196–197
heat transfer coefficient, 14
horsepower for pump, calculation, 89
industrial production of CO_2, 29
integration, 38–39
kinetic energy, 15, 30
linear equation system by Excel, 40–41
linear equation system by MATLAB®, 54–55
lithium hydroxide, calculation, 31
mass flow rates of outlet stream, calculation, 160–161, 197–198
matrix method, 39, 42–43
methane gas, molar volume and density, 17
methanol in reactor, per pass conversion of, 182–183
minimum reflux ratio, 163–164
Newton–Raphson method, 38
NPV, 200
operating pressure, 128–129
partial differential equation, 110–111
physical properties for water, estimation, 116
physical separation process, 30
pipeline diameter, calculating, 89
product component flow rates, calculation, 184–186
radioactive decay of polonium, 42
Raoult's law, 55, 63
shell and tube heat exchanger, 112–113
sodium sulfite, 27–28
solvent recovery system, 31–32
terminal velocity of spherical particle, 91–92
thermodynamic feasibility of chemical reaction, 182
thickener, area and diameter of, 92–94
total pressure, calculating, 15
Txy diagram for ethanol–water system, 59
van der Waals equation, 16
vapor-liquid-equilibrium (VLE) data, 63
water–ethanol mixture, average density of, 43
S200 Omni Processor, 207
Space time, 177–178
Standard atmosphere, 7
Static pressure, 74
Steady state flow, 72
Stoke, 71
Stress, defined, 69–70
Stripping
analytical methods, 154–156
defined, 119, 147
design, 149
in distillation columns, 127, 130
operating line and equilibrium curve for, 150
Synthesis process, 212, 241
Synthesis reaction, 167
Systems of units, 3–5

T

Tangential stress, 70
Tangent method, *see* Newton–Raphson method
Temperature, 95
scales, 6–8
Terminal velocity of spherical particle, solved examples, 91–92
Theoretical approach, xxii
Theoretical component, xxii
Theoretical models, xxii
Theoretical stage, *see* Equilibrium stage
Thermal conductivity, 96–98
Thermal decomposition of NO_2, 23
Thermal radiation, 97, 106
Thermodynamics
defined, 165
entropy, 179–180
equations of state, 178–179
equilibrium concept, 178
free energy of formation of reaction, 180–181
laws, 26, 179
process concept, 178
role of, 22–23, 27
Third law of thermodynamics, 26, 179
Time-varying problems, 47, 241
Total capital investment, 193–194
Total condenser, 125
Total head, 78
Total reflux (TR), 138
Transfer unit concept for absorption, 156
Transient problems, 47, 241
Transparent models, *see* Theoretical models
Transport phenomena, xxiii
role of, xxv–xxvii
Tray efficiency, 138–140
Trial and error procedures, 37
Turbulent flow, 72
Txy diagram
for benzene–toluene mixture using Excel, 60–62, 64
for ethanol–water system, 59

U

Underwood equation, 142–143
Uniform flow, 72
Unit operations, xxiv, 19–20
role of, xxv–xxvii
Unit process, xxiv, 20
role of, xxv–xxvii

Universal constant, *see* Gas constant
Unsteady state flow, 72
U-type heat exchanger, 108

V

Values of gas constants, 236
van der Waals equation, 11
    example, 16
Vapor–liquid equilibrium constant, 147
Vapor pressure, 121
Vapor product, 127
Velocity gradient, 73–74
Viscosity, 71

W

Water pumps, 86
Well pumps, 86

Z

Zero-order reaction, 169, 170
Zeroth law of thermodynamics, 26, 179